Climate and History

Climate and History

Studies in Interdisciplinary History

Edited by
Robert I. Rotberg and Theodore K. Rabb

Contributors
Andrew B. Appleby
Donald G. Baker
Micheline Baulant
Reid A. Bryson
Jan de Vries
John A. Eddy
David Hackett Fischer
Harold C. Fritts
Geoffrey A. Gordon
David Herlihy
Emmanuel Le Roy Ladurie
Helmut E. Landsberg
G. Robert Lofgren
Jerome Namias
Christine Padoch
Christian Pfister
John D. Post
Theodore K. Rabb
Thompson Webb III
Alexander T. Wilson

Princeton University Press
Princeton, New Jersey

Introduction copyright © 1981 by
Princeton University Press
Published by
Princeton University Press,
Princeton, New Jersey
In the United Kingdom:
Princeton University Press,
Guildford, Surrey

LCC 81-47285
ISBN 978-0-691-64241-3

Printed in the United States of America by Princeton University Press, Princeton, New Jersey

The articles included in this reader are reprinted by permission of the authors and the editors of the *Journal of Interdisciplinary History*. They originally appeared in the *Journal* as indicated below:
Brysen and Padoch, "On the Climates of History," *JIH* X (1980), 583-597: copyright © the JIH & MIT
de Vries, "Measuring the Impact of Climate on History: The Search for Appropriate Methodologies," X (1980), 599-630: copyright © the JIH & MIT
Landsberg, "Past Climates from Unexploited Written Sources," X (1980), 631-642: copyright © the JIH & MIT
Appleby, "Epidemics and Famine in the Little Ice Age," X (1980), 643-663: copyright © the JIH & MIT
Pfister, "The Little Ice Age: Thermal and Wetness Indices for Central Europe," X (1980), 665-696: copyright © the JIH & MIT
Namias, "Severe Drought and Recent History," X (1980), 697-712: copyright © the JIH & MIT
Herlihy, "Climate and Documentary Sources: A Comment," X (1980), 713-717: copyright © the JIH & MIT
Eddy, "Climate and the Role of the Sun," X (1980), 725-747: copyright © the JIH & MIT
Webb, "The Reconstruction of Climatic Sequences from Botanical Data," X (1980), 749-772: copyright © the JIH & MIT
Fritts, Lofgren, and Gordon, "Past Climate Reconstructed from Tree Rings," X (1980), 773-793: copyright © the JIH & MIT
Wilson, "Isotope Evidence for Past Climatic and Environmental Change," X (1980), 795-812: copyright © the JIH & MIT
Baker, "Botanical and Chemical Evidence of Climatic Change: A Comment," X (1980), 813-819: copyright © the JIH & MIT
Fischer, "Climate and History: Priorities for Research," X (1980), 821-830: copyright © the JIH & MIT

FIRST PRINCETON PAPERBACK, 1981

Contents

Introduction

The study of past climates is an old pursuit. Helmut Landsberg, in the third chapter of this book, provides a clear picture of early methods of recording and examining the world's climate. But collaborative research on the impact of climate on man's works and events, and a systematic reporting and evaluation of that research were uncommon before the *Journal of Interdisciplinary History*, with support from the Rockefeller Foundation, sponsored a meeting of meteorologists, paleobotanists, astronomers, chemists, physicists, and historians in 1979. The results of that conference, suitably revised, appear here in book form for the first time. (They originally appeared in the tenth volume of the *Journal of Interdisciplinary History*, Spring, 1980.)

The following chapters suggest ways in which historians and climatologists should collaborate in order to appreciate the place of climate in man's past. A number of the essays are methodological: Harold Fritts and his colleagues explain how tree-ring dating can be controlled and used for the purposes of history. Thompson Webb III provides an unusually sophisticated series of suggestions for the reconstruction of climate patterns from preserved pollen data in ice cores and lake sediments. John Eddy speculates on the role of the sun in explaining the climatic sequences of the past. Alexander Wilson shows that the chemical and isotope composition of strategraphic deposits—in caves, oceans, tree rings, and ice sheets—can provide detailed information about the nature of man's environment for periods that preceded the historian's written records. Although man's ability to predict drought has been limited, Jerome Namias' article examines physical factors associated with drought and attempts to trace global patterns in the circulation of the atmosphere. David Herlihy, a historian of Medieval and Renaissance Italy, shows how documentary sources long familiar to historians can be used to develop climatic data of relevance both to historians and to climatologists.

Reid Bryson and Christine Padoch demonstrate why proxy data must be employed to provide accurate historical readings. In his article, Bryson, a pioneering climatologist, distills the experience of many decades during which he has sought to ensure appropriate collaboration between scientists of climate and historians conscious of the contribution of climate.

Christian Pfister and Emmanuel Le Roy Ladurie, authors of two contributions to this volume, have been in the vanguard of those demonstrating how European proxy data that pertain particularly to harvests and wine pressings can (and cannot) be employed to enrich our appreciation of the impact of climate on the past. Barbara Bell, an astronomer, proposes a rigorous mathematical method for reconciling varieties of viticultural data in order to discern underlying information about rainfall and mean temperatures over time.

Jan de Vries provides a critique of several important methodologies; he comes to the controversial conclusion that retrospective measurement and approximation may only inform us about short-term problems. The long-run impact of climate on history should, he argues, be a record of man's ability to adapt. He recommends a history of accomplishment as influenced by climate, not a history of crises. Donald Baker agrees, stressing the biosphere's inherent resilience—an overall quality which may contribute to serious distortions of the historical record.

Andrew Appleby's chapter focuses upon the role of climate and climatic change in the morbidity and mortality of Europeans, especially during the little ice age. As readers of this entire book will see, however, the contributors to this volume now doubt that historians should talk confidently of a little ice age. There may either have been a succession of cold decades interspersed with warmer ones, or real climatic phenomena of the period may (as de Vries and Appleby indicate) have had too little demonstrable impact on history to merit the title, "Ice Age." Many of the comments included in this book reflect the diverse views of their authors on this question—now a central one in historiography.

This volume concludes with two essays (by David Hackett Fischer and Theodore K. Rabb) that set out priorities for future research between historians and scientists of climate. The editors intend the publication of this book to stimulate the kinds of mutual development which will make joint research on the role of climate in history productive and central to several disciplines.

—R.I.R., T.K.R.

Climate and History

Reid A. Bryson and Christine Padoch

On the Climates of History

The development of objective, quantitative evidence of how climates or *climata,* and the associated *biota,* have changed significantly (even during post-glacial and historical times) has expanded the possibility of the rational inclusion of the climatic factor in the study of history.[1] Climatic variation has produced variation in both the quantitative and qualitative character of the economic base of cultures, nations, and societies. This new recognition is not a revival of environmental determinism; it implies neither that all environmental changes have a climatic cause, nor that all cultural changes have an environmental cause, and it does not rest on an assumption that the links between climatic and human history are simple or straightforward. This new appreciation of the role of climates and climatic change is rather an extension of well-known ecological principles.

One of these principles predicts that changes in community composition will result from shifts in relative competitive advantage when environmental factors change. In the human context, this principle suggests that the physical environment, and particularly the climate, gives a bias to the direction and success of the near-infinite series of decisions that make up the course of history. That is, with a shift in temperatures, or of amounts and timing of rainfall, the particular mix of resource use techniques characteristic of a population may well change, some occupations supplanting others as they become more profitable, less risky, and therefore more important. Clearly the size and rapidity of climatic change is crucial to the ability of societies to adapt or to cope.

Another of these principles is that of limiting factors. To use an example from limnology, nitrogen may be limiting in one lake, phosphorus in another. Lack of heat is limiting in the Arctic,

Reid A. Bryson is Director of the Institute for Environmental Studies at the University of Wisconsin, Madison. He is the author with Thomas Murray of *Climates of Hunger* (Madison, 1977). Christine Padoch is Assistant Professor at the Institute for Environmental Studies at the University of Wisconsin, Madison.

1 The term *climata* is used here to indicate coherent assemblages of climatic elements in the same sense that the term *biota* is used to indicate natural assemblages of biological elements.

lack of water in the desert. Climatic fluctuations may move temperature, precipitation, or other climatic thresholds and change the absolute limits of particular economic activities, altering previous patterns. New possibilities may also be opened or old patterns eliminated.

What then do we know about the climates of history?

DOCUMENTARY EVIDENCE There are scattered references to climatic change, and its impact on human activities, that go back to Greek and Roman times. Cyprian, about 250 A.D., commented on the diminution of winter rains and summer heat in Tunisia. Less clear is Avienus' description of the deserted and desert nature of the east coast of Spain in the fourth century A.D. It is not until the eighth or ninth century, however, that the documentary evidence becomes abundant enough to make possible a nearly continuous account for some regions.[2]

As recently as 1962, reconstruction of past climates was largely qualitative, or at best semi-quantitative. The relative abundance of comments on the severity or mildness of winters, as gleaned from such sources as chronicles and diaries, gave a rough measure of the character of a decade. Similar relative measures of summer wetness were possible. More quantitative but less direct were the records of grain prices, numbers of prayers said for rain per year, duration and extent of sea ice, outbreaks of weather-related disease, famines, and the like. Interpretation and calibration were necessary, and one was never sure whether some of the variation from one year to the next might not be due to non-climatic economic or political factors.

At a conference on the climates of the eleventh and sixteenth centuries, the climatologists were surprised by, and enthusiastic about, the wealth of quantitative data on climate-related parameters that historians could provide for a reconstruction of climate, but none knew exactly how to interpret the data, even when

2 Quotations from Cyprian and Avienus may be found in Rhys Carpenter's *Discontinuity in Greek Civilization* (New York, 1968), 82–85. Gordon Manley, "Temperature Trends in England, 1698–1957," *Archiv für Meteorologie, Geophysik, und Bioklimatologie*, IX (1959), 412–433; Hubert H. Lamb, *The Changing Climate* (London, 1966); Emmanuel Le Roy Ladurie (trans. Barbara Bray), *Times of Feast, Times of Famine: A History of Climate since the Year 1000* (Garden City, 1971).

gathered by the most skillful historians. The historians, on the other hand, wanted to know what the climates had been in order to interpret history.[3]

A primary problem with the use of documentary evidence (other than documented instrumental data) in the reconstruction of past climates is the problem of compound parameters. Most of the documented data are derived from such items as grain prices, which depend on a variety of factors, both economic and climatic (and *their* interaction). The reconstruction of the variations of individual factors, such as summer temperature or rainfall, require as many different sets of data as there are interacting factors. This requirement is simply a restatement of the rule that the simultaneous solution of a set of linear equations requires as many equations as there are unknown quantities.

A second problem is that of critical lacunae. In a discussion of evidence of drought at the time of the disappearance of a particular culture, a palynologist once argued that a core from a nearby lake contained no evidence, in the pollen record, of drought at that time. When the original paper on the pollen analysis of that core was checked, it was found that the lake had totally dried up during that period and that there was thus no pollen *evidence* because there was *no pollen* preserved. It is likely that this problem may also be a function of historical records. The "times of troubles" have more fragmentary records than stable times.

Another problem arises with subjective records of climate such as "this has been the coldest winter in memory"—the data filter of human recollection. Such references are to the perceived normal and thus might record short-term, but probably not long-term changes. For example, Lamb's analysis of the chronicles from Russia shows a great variation in the number of severe winters prior to the time of Napoleon's invasion, but not after. Did the disappearance of severe winters really occur? That is hardly consistent with the rest of the European data, since severe winters in Western Europe are associated with the flow of very cold air out of Russia and central Asia. Or was there a psycho-

3 Reid Bryson and Paul Julian (eds.), *Proceedings of the Conference on the Climate of the Eleventh and Sixteenth Centuries* (Boulder, 1963).

logical change, a severe winter being nothing to the Russians but intolerable to the invaders—a new pride that changed the perception of severe winters?[4]

The resolution of these problems requires objective, quantitative data on climate which are independent of historical data. Although great progress has been made using documentary evidence in understanding the sequence of past climates, such parallel climatic data would eliminate interpretive problems, provide a calibration for proxy data, and open up new possibilities in the separation of environmental from cultural factors. Fortunately, there are some new developments in paleoclimatic science that should foster this deeper historic insight, especially for cultures without a written history.[5]

OBJECTIVE, QUANTITATIVE CLIMATIC RECONSTRUCTIONS Climate varies on many scales of space and time. Interannual variations are usually highly correlated over short distances, but the correlation decreases with distance so that at some distance (often hundreds of kilometers) the correlation becomes negative. At still larger distances the correlation may again be positive and significant. These far-off positive correlations are often called teleconnections. The same is true of decadal and probably century-long variations. This situation is fundamentally a logical consequence of the atmosphere being a dynamic, interconnected whole, the large-scale dynamics requiring, through changes in the atmospheric wave structure, that some regions warm when others cool and some become wetter when others become drier. Precipitation often varies over shorter time and distance scales than does temperature.

There may also be variation in the rate of change. Some climatic changes appear to be rapid steps from one climatic state

4 As shown in John Kutzbach and Bryson, "Variance Spectrum of Holocene Climatic Fluctuations in the North Atlantic Sector," *Journal of the Atmospheric Sciences,* XXXI (1974), 1958–1963, Figure 1, there is less variance in documentary records at periods greater than about thirty-five years than there is in objective or instrumental records. For shorter periods the variance is about the same. This does not mean, necessarily, that the documentary records are correct. Lamb, *Changing Climate,* 219–221.

5 See for example, the transfer-function method of converting quantitative biological data to climatic data in Thompson Webb III and Bryson, "The Late- and Post-Glacial Sequence of Climatic Events in Wisconsin and East-Central Minnesota: Quantitative Estimates Derived from Fossil Pollen Spectra by Multivariate Statistical Analysis," *Quaternary Research,* II (1974), 70–115.

to another in some regions, especially those near climatic quasi-discontinuities (e.g., mean frontal positions), whereas others, far from such boundaries and tempered by maritime influences, appear to have more gradual climatic transitions.

There is no true substitute for an instrumental record of the climate at the place of interest to the historian. However, if the time of interest is more than a century into the past, the probability of such a record existing diminishes rapidly to zero. For the reasons mentioned above one cannot, even for the seventeenth through the nineteenth centuries, use whatever record happens to be available *somewhere*. The central England temperature record for the past few centuries is representative of central England, less so of Western Europe, still less so of Scandinavia, etc. It certainly must be regarded with some skepticism as being representative of the hemisphere or globe unless there is corroborative evidence.[6]

Fortunately, several methods have been developed for deriving objective, quantitative climatic data from proxy data. Most of them use biological data, although glaciological and chemical data may be useful in some areas. Unfortunately, year by year and month by month detail is not generally possible, even with the use of tree rings and annual laminae in sediments and glaciers. Trees blur several years together because of various storage mechanisms and respond to several parameters of the climate. For instance, if a drought has proceeded long enough for the weaker trees in an area to succumb, the roots of the surviving trees may have access to a larger area of soil, and thus to more water, and the tree may receive more light. The surviving trees may grow better, simulating more rainfall, despite a continuation of the drought.

However, even proxy records that are less than perfect and give decadal to century estimates have their use. Such records from the past few centuries, if they were taken closer to the historian's region of interest than the nearest instrumental record, may reflect the climate of concern as well or better. For earlier times they are essentially the only data.

6 Reconstructions from a variety of proxy records show that the central England temperature record is *similar* to the records from the Great Lakes area, Iceland, and northern Finland, and these regions in turn are indicative of the temperature in the far-north Atlantic and the eastern Canada area. This region in turn has provided a large part of the variance of the hemispheric mean temperature observed in the past century.

A prime use of climatic data in historical analysis is to assess the degree of stress on the economic base of the people being studied; this is often an agricultural or fisheries base. For this purpose the state of the art in climatic reconstruction is sufficiently far advanced for considerable progress to be made, especially for cultures with no, or minimal, documentation. Even where the local conditions or chance result in no climatic reconstruction having been made as yet, the coherent structure of the atmosphere can provide a means of estimating the climate using data from far distant places with at least as much confidence as one should have in the use of an actual record from a single place a thousand kilometers away.[7]

From the existing proxy records of environmental variation, primarily reflecting climatic variation, one may reach two conclusions:

1. It is not necessary to theorize about what the climate might have been on the basis of some putative *cause* of climatic change, at least for the past few centuries. There are adequate tools for assessing the proximal cause of environmental change without resorting to ultimate causes and theory, much of which is still untested against reality.

2. There have been times of globally synchronous, rapid, climatic (environmental) change. These should have been times of maximum stress for some cultures and improving times for others.

A TIME FRAMEWORK OF HISTORICAL ENVIRONMENTAL CHANGE In the context of the present discussion one can at least give a framework of times of apparent rapid change between climatic regimes. This framework does not provide the detailed local sequence of specific climatic data that one might wish to have for studies of local short-period history, but it does provide important background for the analysis of historical trends and developments.

Climatic data are often expressed in terms of means of temperature or totals of precipitation. That a decade was a degree or

7 An example of using data from one region to reconstruct climates in far distant regions is found in Harold Fritts, T. J. Blasing, B. P. Hayden, and Kutzbach, "Multivariate Techniques for Specifying Tree-Growth and Climate Relationships for Reconstructing Anomalies in Paleoclimate," *Journal of Applied Meteorology*, X (1971), 845–864. Some unreported experiments suggest that a similar approach might make possible a general coverage of Europe from the few yearly climatic series available since 1600 A.D. or so.

two colder than the preceding one does not sound impressive—especially to those of us who know that if we change the temperatures in our houses by a degree or two by changing the setting of the thermostat we really do not notice a significant difference. However, an often overlooked fact about climatic data is that small differences in the mean temperature may mean significant differences in the frequency of occurrence of extreme values. Even in the midwestern United States, a change of mean July temperature of 2–3°F may be associated with a change in frequency of the extremes which are stressful to crops by a factor of five to ten. In Iceland a decrease of annual temperature of 1°C reduces the growing degree days by 27 percent, illustrating that small changes may be critical in marginal areas.[8]

If we restrict our attention to past cultures, as recognized by the scholars who study them and objectively date them, there are two significant dates—the endpoints. Intermediate dates may be regarded as a sampling from the cultural continuum. The earliest and latest dates that a culture would actually be recognizable are not generally known, however, since the probability of those points having been dated is small. However, we can treat the 150 or so radiocarbon dated cultures as a set of continua with only approximately known termini, and apply statistical tests to see whether there are significant times marked by global cultural change. The results of such a study show that there have been highly significant times of culture change, some being nearly ubiquitous. The significant dates from that study, and a later one appear in Table 1.[9]

It should not be concluded from these dates that every society would have been experiencing climatic problems at those times, for nearly all climatic changes about which much is known have been such that some areas got wetter, some drier, some warmer, some colder, *and some showed very little change*. It cannot be assumed that, because some areas showed evidence of climatic change that conflicted with some other area, and some areas showed no change, the evidence is wrong. Furthermore,

8 Bryson, "Airmasses, Streamlines, and the Boreal Forest," *Geographical Bulletin*, VIII (1966), 228–269, Table 2; *idem*, "A Perspective on Climatic Change," *Science*, CLXXXIV (1974), 753–760.

9 Wayne M. Wendland and Bryson, "Dating Climatic Episodes of the Holocene," *Quaternary Research*, IV (1974), 9–24; Bryson, "Cultural, Economic and Climatic Records," in A. B. Pittock et al. (eds.), *Climatic Change and Variability—A Southern Perspective* (Cambridge, Mass., 1978), 318.

Table 1 Partial List of Dates of Globally Synchronous Maxima of Culture Change Rates.[a]

CHI-SQUARED METHOD		CUMULATIVE CHANGE METHOD	
MORE SIGNIFICANT	LESS SIGNIFICANT	CLEAR	LESS CLEAR
—	1260–1290 A.D.	1340 A.D.	—
1150–1170 A.D.	—	1100–1130 A.D.	—
—	890 A.D.	—	890 A.D.
690 A.D.	—	660 A.D.	—
—	145 A.D.	160 A.D.	—
775 B.C.	—	410 B.C.	—
1480 B.C.	—	1460 B.C.	—
—	2110 B.C.	—	2110 B.C.
2950–2970 B.C.	—	2990–3100 B.C.	—
—	4415 B.C.	—	4410 B.C.

a The culture change rates are statistically derived from radiocarbon dated continua, converted to calendar date using tree-ring calibration. Calibration according to E. K. Ralph, H. N. Mitchell, and M. C. Han, "Radiocarbon Dates and Reality," *MASCA Newsletter*, IX (1973), 1–20.

there might have been non-environmental factors operating on a global scale, although it is difficult to imagine what, other than geophysical processes, might have operated synchronously or at least within a century or two over most of the globe in prehistoric times.

An additional study was done to determine which of these times significant to culture history might have been times of environmental change. All dates reported in *Radiocarbon* as indicating times of environmental change were extracted, and statistically analyzed. There are hundreds of such dates, which tend to cluster at preferred times (Table 2). Work not yet completed suggests that the clustering is even sharper if the dates reported in *Radiocarbon* are adjusted to a uniform set of criteria as to data reduction and choice of beginning date for a change (rather than some being beginning, some middle, some previous peak, etc.).[10]

The dates in Tables 1 and 2 do not match very well but, given the uncertainty of radiocarbon dating and data selection,

10 Wendland and Bryson, "Dating Climatic Episodes." The thousands of radiocarbon dates in the twenty-one volumes of the journal *Radiocarbon* constitute as a whole a body of data which has only rarely been treated as a statistical assemblage.

Table 2 Partial List of Globally Preferred Dates of Climatic Change.[a]

WENDLAND AND BRYSON[b]		FRENZEL[c]
MAJOR	MINOR	
—	—	1405 A.D.
1110–1130 A.D.	—	—
280 A.D.	—	390 A.D.
940–990 B.C.	—	1020–1050 B.C.
—	2080 B.C.	2110 B.C.
—	2940 B.C.	2890–2900 B.C.
2830–3850 B.C.	—	3710 B.C.
—	5000 B.C.	4910 B.C.

a The dates of climatic change are determined from biological and geological indicators, converted to calendar dates using tree-ring calibration. Calibration according to Ralph et al., "Radiocarbon Dates."

b Wendland and Bryson, "Dating Climatic Episodes."

c Frenzel, "The Distribution Pattern of Holocene Climatic Change in the Northern Hemisphere," *Proceedings of WMO/IAMAP Symposium on Long-Term Climatic Fluctuations* (Geneva, 1975), 105–118.

many are close enough to each other that the scholars of those periods should consider the hypothesis of climatic change as a factor in cultural change and historical events at those times. Indeed, it would be wise to consider the hypothesis for any time of global cultural change because geophysical changes have been more tightly linked globally than have human events.

There are more times of known climatic changes than are indicated in Table 2. Various proxy data series indicate, at least for Europe and North America, that important changes occurred around 1000 A.D., 1150–1200 A.D., c. 1450 A.D., and in the 1550–1600 A.D. period.

Summing up the studies mentioned above and many more, we may construct a tentative global sequence of climatic episodes, rather similar to the geological framework of epochs separated by significant changes or "revolutions" (Table 3). The names given come primarily from the Blytt-Sernander European sequence derived from palynological studies, since that time framework appears to be globally applicable. The post sub-Atlantic sub-episode names are taken from other work and are deemed preferable to

Table 3 Tentative Division of the Holocene into Climatic Episodes, Based on Globally Preferred Dates of Climatic Change.[a]

CLIMATIC EPISODE	SUB-EPISODE	PROVISIONAL TERMINI	POSSIBLE SUBDIVISION TERMINI	CHARACTER	
		Present			
	Modern			Maximum warmth 1945± (mean n. hemisphere)	
		1915–1920 A.D.	1915 A.D.	Cool n. hemisphere	Little Ice Age
	Neo-Boreal		1885	Mild n. hemisphere	
Post Sub-Atlantic			1820	Generally cold n. hemisphere	
			1765	Mild n. hemisphere	
			1720	Coldest 1600–1630, 1670–1720 n. hemisphere	
			1600	Rapid drop of temperature, esp. N. Atlantic area	
		1550 A.D.		Warm N. Atlantic	
	Pacific		1400	Cooler N. America-N. Atlantic	
		1150–1200 A.D.			
	Neo-Atlantic		1000	Medieval warm period	
		700–750 A.D.			
	Scandic			Character largely unknown	
		300–400 A.D.			
Sub-Atlantic	S-A III				
		50–100 B.C.		Beginning of 2,000-yr. general decline of N. America summer temperatures	
	S-A II				
		ca 500 B.C.		Highly unstable, weak monsoon NW India, frequent drought in China	
	S-A I				

		ca 950 B.C.		
Sub-Boreal	S-B II		Monsoon failure NW India, expanded tundra, glacial advances	
		2100 B.C.		
	S-B I			
		2900–3000 B.C.		
Atlantic	A IV		Probably warmest post-glacial epoch	Climatic Optimum / Altithermal / Hypsithermal (A IV–A I)
		c 4920 B.C. (5970 BP ^{14}C)		
	A III		Cooler than AI and AIV (c 4920 B.C.–7060 BP)	
		6740 BP (^{14}C)		
	A II			
		7060 BP (^{14}C)		
	A I		Quite warm	
		7900 BP (^{14}C)		
Boreal	B II		Cochrane glacial advance	
		8490 BC (^{14}C)		
	B I			
		c 9160 BP (^{14}C)		
Pre-Boreal			Monsoons expand dramatically, rapid warming	
		10800 BP (^{14}C)		

a The nomenclature is modified from the Blytt-Sernander scheme.

such loosely defined terms as "little ice age." After all, glacial advances did not occur in all parts of the world.[11]

It is not possible to summarize in a table all the detail that is known about each of the climatic episodes, and it is not possible, in general, to characterize even sub-episodes as "cold," for example. In almost every case data can be found to show that there was a varied direction of change, even though the times of change were globally common. Some examples will be outlined in the following section.

CASE STUDIES A good example of one use of climatic data to elucidate a historical question is the work that has now been done on the decline of Mycenae and other Late Bronze Age civilization patterns in the Near East, around 1200–1000 B.C.

Carpenter had proposed, on the basis of a life-long study of Mycenae, that there was little evidence to support the long-held assumption that a Dorian invaion had led to the decline of Mycenae. He proposed instead that a long period of drought had so disabled the Mycenaean agricultural base that the people had sought greener pastures. His problem with the drought hypothesis was that there would have to have been a very patchy pattern of drought in Greece and the neighboring lands, for although Crete, the southern Peloponnesus, Boeotia, Phokis, and the Argolid presented evidence of decline, Attica, the western slopes of the Panachaic Mountains, Kephallenia, Thessaly and northern Greece, and Rhodes did not. He attempted to rationalize that such a pattern of irregular drought distribution was possible in a topographically irregular region.[12]

That such a pattern is not only possible but is a normal mode of rainfall and drought variation was later demonstrated. Using modern data, Donley calculated the eigenvectors of the winter

11 David Baerreis and Bryson, "Climatic Episodes and the Dating of the Mississippian Cultures," *Wisconsin Archaeologist*, XLVI (1965), 203–220. The Blytt-Sernander terminology is widely used but rarely referenced. A. G. Blytt, *Essays on the Immigration of the Norwegian Flora during Alternating Rainy and Dry Periods* (Oslo, 1876) provided a basis, modified by Rutger Sernander, "Die schwedischen Torfmoore als Zengen postglazialer Klimaschwankungen," in *Die Veränderungen des Klimas seit dem Maximum der letzten Eiszeit* (Stockholm, 1910), 197–246. Neither could attach objective dates to their pollen zones, of course.

12 This explanation of the decline of Mycenae was the main thrust of Carpenter's *Discontinuity in Greek Civilization*.

precipitation and of a drought index. Although the pattern proposed by Carpenter was not the dominant modern mode of variation, there is no reason why the fifth most common mode at present might not be the dominant mode in some other climatic episode. Indeed there is good evidence of such a change of modes from the neo-Boreal sub-episode to the modern period. The answer to Carpenter's query is clearly that his proposed drought pattern is possible.[13]

Whether it is possible or not, one must then ask whether it actually happened. Based on Donley's work, Bryson, Lamb, and Donley showed that, although it could not be conclusively proven that Carpenter's proposed pattern did occur at the time of Mycenaean decline, the pattern was consistent with other climatic conditions around the world. The dynamic unity of the atmosphere requires consistency, and the world-wide climatic conditions that are associated with the proposed Mycenaean drought pattern at present appear to have prevailed at the time of Mycenaean decline.[14]

Given the present uncertainties of radiocarbon dating, even when calibrated against absolute tree-ring dates, it may be necessary in the light of future research to push the sub-Boreal to sub-Atlantic transition back a couple of centuries.

Another example within the period of more abundant written history is the set of climatic changes at the neo-Atlantic to Pacific sub-episode transition around 1150–1200 A.D. That large-scale climatic changes were taking place in the northern hemisphere, at least, is essentially incontrovertible. The tundra of North America expanded southward rather quickly along a thousand mile front. The extent of sea ice around Greenland and Iceland expanded dramatically. Summer temperatures shown by pollen in annually layered lakes in northern Michigan dropped rapidly,

13 David L. Donley, "Analysis of the Winter Climatic Pattern at the Time of the Mycenaean Decline," unpub. Ph.D. diss. (Univ. of Wisconsin, Madison, 1971). An example of a change in dominant climatic mode at the time of climatic change is found in Blasing and Fritts, "Reconstructing Past Climatic Anomalies in the North Pacific and Western North America from Tree-Ring Data," *Quaternary Research*, VI (1976), 563–580.

14 Bryson, Lamb, and Donley, "Drought and the Decline of Mycenae," *Antiquity*, XLVIII (1974), 46–50. An extension of the drought hypothesis to the whole of Near Eastern history from 1300 to 1000 B.C has been made by Barry Weiss in "The Decline of Late Bronze Age Civilization as a Possible Response to Climatic Change," unpub. ms. (National Center for Atmospheric Research, 1979).

and the dry "rain-shadow" area over the northern plains of the United States expanded eastward. The vegetation of western Iowa changed rapidly from tall grass prairie toward steppe and gallery forests along the rivers diminished. The frequency of wet summers in Eastern Europe and mild winters in Western Europe increased.[15]

All of these changes suggest a southward shift of the baroclinic zone (transition from high-latitude cold air masses to warmer air) and an expansion of the circumpolar vortex about 1150–1200 A.D. Reconstructing from modern data the change in July rainfall that would accompany an expanded polar vortex, i.e., more westerlies between 35°N and 55°N, Baerreis and Bryson tested this hypothesis against archaeological, palynological, and paleozoological data from a variety of locations in mid-North America to see whether the hypothesized pattern of rainfall was correct and what the human impact might have been. Without putting the above data and the investigation into the actual time sequence of research, the results are:

1. There was a dramatic change in climate starting about 1150–1200 A.D., reducing the average July rainfall in the northern plains by perhaps 30–50+ percent.

2. The drier climate lasted sporadically for approximately two centuries.

3. The southern plains became wetter in July.

4. On the northern high plains, which are generally dry, thousands of small villages characterized by rainfed maize agri-

15 The tundra expansion is dramatically recorded in fossil soil profiles of northern Canada, which are tundra soils overlying forest soils. The time of transition can be dated by using the buried forest materials at the interface. The original work was reported in Bryson, W. N. Irving, and J. A. Larsen, "Radiocarbon and Soils Evidence of Former Forest in the Southern Canadian Tundra," *Science,* CXXXXVII (1965), 46–48. The study was extended to cover a much larger area in Curtis J. Sorenson, J. C. Knox, Larsen, and Bryson, "Paleosols and the Forest Border in Keewatin, N.W.T.," *Quaternary Research,* I (1971), 468–473. Other studies by Sorenson and Knox extend the area into Mackenzie District. During the course of the original work Irving, an archeologist, excavated some multiple soil horizons, such as the exposure at Black Fly Cove, Ennadai Lake, N.W.T., and discovered that artifacts of southern or Indian affinities were found in the forest soils, whereas northern or Eskimo affinity artifacts were found in the tundra soils. Evidently the historic association of Chipewyan with the forest and Caribou Eskimo with the tundra has a 4,000 year basis.

Pall Bergthorsson, "An Estimate of Drift Ice and Temperature in Iceland in 1000 Years," *Jokull,* XIX (1969), 94–101; Albert Swain, "Environmental Changes during the Past 2,000 Years in North Central Wisconsin: Analysis of Pollen, Charcoal, and Seeds from Varved Lake Sediments," *Quaternary Research,* X (1978), 55–68.

culture before 1100 A.D. had completely disappeared by 1200 A.D. and many became covered with wind-drifted soil.

5. Farther east, in western Iowa, the Mill Creek culture (of maize farmers) had occupied valley floors with forested terraces in a region of generally tall-grass prairie. After 1200 A.D. the gallery forests were gone and short-grass prairie dominated. About 1150–1200 A.D. the meat diet of the Mill Creek people changed from dominantly deer (a forest browser) to dominantly bison (a short-grass grazer). After a little less than two centuries the culture succumbed.

6. After 1200 A.D. a number of villages practicing rain-fed maize agriculture were established in the Panhandle region of Texas and Oklahoma, in a region where that had not been possible before and is not at present.

The climatic change of 1200 A.D. had completely changed the occupation pattern and subsistence base of a region comprising a minimum of 900,000 square kilometers. Other evidence suggests that much of the inhabited world was affected in one way or another.[16]

THE NEW COLLABORATION There is a body of paleoclimatic techniques capable at present of producing broad background information on the climates of almost any period in the history of man. A body of climatic theory exists that makes the cross-checking and inter-relating of climatic data possible as well as a collection of ecological principles which can be applied to the man-climate relationship. There are indications that research in the near future will provide climatic data which are much more specific in time and place although probably never with the detail of an instrumental record.

Several conferences and workshops have now expanded the dialogue between historians and paleoclimatologists. When historians learn to use what the climatologists can provide, and the climatologists focus their efforts on areas and times of particular interest rather than on data sets and places which are basically "targets of opportunity," a new collaboration may emerge that will help to answer the *why* as well as the *what* of human history.

16 An account of this and related research, including the Baerreis and Bryson work, is given in Bryson and Murray, *Climates of Hunger* (Madison, 1977).

Jan de Vries

Measuring the Impact of Climate on History: The Search for Appropriate Methodologies

A generation ago the proposition that a non-human history systematically impinged upon, let alone shaped, human history found little support among historians. Today, the acceptance of this possibility is regularly being urged upon us, and by many of our most distinguished colleagues. Yet, to what extent is this unmistakable intellectual shift the result of convincing evidence of linkages between climate and human history of more than a fleeting and random sort?

This article examines the measurement of climate's impact on economic life in preindustrial Europe in theory and practice. I address, in turn, climatic influences on the three levels of historical time—short-term, conjunctural, and long-term—with the purpose of assessing methodologies and identifying results.

DATA Historians who would seek to examine the impact of climate on economic life are immediately confronted by the treacherous fact that climatological data are both extremely limited and all-pervasive. Extremely limited are instrumental measurements of meteorological phenomena. Before the mid-eighteenth century we are forced to rely on the central English series of Manley and the Dutch series of Labrijn, and these take us back no earlier than the late seventeenth century. But, once the investigator, seeking to extend his knowledge to earlier periods, begins to use non-instrumental data he faces another sort of problem: potential sources abound but their interpretation is complex.[1]

The sources of climate evidence and their problems are dealt with elsewhere, but here I wish to call attention to two matters critical to any study of the influence of climate on economic life. Evidence drawn from scattered observations and from the casual

Jan de Vries is Professor of History at the University of California at Berkeley. He is author of *The Dutch Rural Economy in the Golden Age* (New Haven, 1974).

1 A. Labrijn, "Het klimaat van Nederland gedurende de laatste twee en een halve eeuw," *Koninklijk Nederlandsch Meteorologisch Instituut, Mededeelingen en Verhandelingen*, 49 (1945); Gerald Manley, "Central England Temperatures: Monthly Means from 1659 to 1973," *Quarterly Journal of the Royal Meteorological Society*, C (1974), 389–405.

comments of contemporaries (such as chronicles, diaries, and historical narratives) suffer from the well-known weakness emphasized by Le Roy Ladurie in his criticism of Utterström's pioneer article.[2] At the heart of our enterprise is measurement, and this requires data which are, in Le Roy Ladurie's words, "continuous, quantitative, and homogeneous." Notwithstanding the brave attempts of scholars such as Easton to pin numbers to the qualitative comments of contemporaries, temporally discrete and anecdotal evidence cannot serve the purposes under discussion here.[3]

"Proxies," a second sort of evidence, are time series of nonmeteorological events that possess the characteristic of sensitively reflecting meteorological phenomena. Examples of widely-used proxies include phenological data (such as the beginning of the wine harvest or the blossoming of the cherry trees in Japan), dendroclimatological data (the measurement of tree rings), and the observation of the dates during which bodies of water are ice-covered.

Such proxy data can meet the objections leveled against the use of scattered observations, but the user must guard against a second pitfall, to which the first category of evidence is also liable, of measuring economic (or social or political) phenomena and attributing a meteorological value to them. That is, although proxy data are continuous and quantitative, they might not be homogeneous. For the climatologist this lack of homogeneity results in poor estimates of meteorological values; for historians seeking to uncover the impact of climate on human society it results in a classic example of circular reasoning. The traditional skepticism of many economic historians toward the claims of climate history arises from this danger. It is not so much that

2 For sources of climate evidence see the contributions of Emmanuel Le Roy Ladurie, Christian Pfister, and Helmut Landsberg elsewhere in this issue. Le Roy Ladurie, "Histoire et climat," *Annales*, XIV (1959), 3–34, translated as "History and Climate," in Peter Burke (ed.), *Economy and Society in Early Modern Europe* (New York, 1972), 134–169; Gustaf Utterström, "Climatic Fluctuations and Population Problems in Early Modern History," *Scandinavian Economic History Review*, III (1955), 3–47.

3 Charles Easton, *Les hivers dans l'Europe occidentale* (Leiden, 1928); for a more refined attempt see Pierre Alexandre, *Le climat au Moyen Age en Belgique et dans les regions voisins* (Louvain, 1976). An exception to this generalization could arise when there exist many independent observations of the same phenomenon. As a practical matter this seems very unlikely to occur.

economic historians believe that they have better explanations for such events as the disappearance of grain cultivation from Iceland, of viticulture from England, or the extension of olive cultivation in France, but that in the absence of independent climatic evidence, such "events" cannot be made to serve as both meteorological evidence and evidence of the human impact of climate.

To test propositions about the human impact of climate, time series are needed, and proxies need to be assigned specific meteorological values. That is, we need quantitative answers to such questions as: what is the likely meaning of Japanese cherry trees blossoming on April 10 rather than April 20; what does a fifteen-day delay in the declaration of the *vendange* imply about the preceding summer. To answer simply that later wine harvests or tree blossomings mean "colder" preceding seasons is unsatisfactory. Statistical comparisons of such data with instrument-measured meteorological data when it becomes available is desirable: methods such as regression analysis not only can supply a specific answer to the questions posed above, but they force the investigator to specify the other factors that may influence the proxy, and help establish the limits of a proxy's usefulness.[4]

ANNUAL FLUCTUATIONS No one is likely to dispute that weather fluctuations affect year-to-year economic conditions, and that in the preindustrial society they did this primarily through the harvests. Precisely because the short-term impact of climate seems so obvious several hardy generalizations have frequently substituted for careful analysis. Widely accepted are two assertions. One is that climatic fluctuations dominated preindustrial society in a direct way but gradually receded in importance in the face of nineteenth-century industrialization. The second is that the economic history of *ancien régime* Europe is essentially the summation of its harvests, which, in turn, were sensitive registers of its climate history, understood as a series of exogenous movements occurring randomly around a long-term mean. In *Capitalism and Material Life,* Braudel expressed such a view: "The world between the fifteenth and eighteenth centuries consisted of one vast peas-

4 For a more extended discussion of this point see: de Vries, "Histoire du climat et économie: des faits nouveaux, une interprétation différente," *Annales,* XXXII (1977), 202–207.

antry where between 80 percent and 95 percent of people lived from the land and from nothing else. The rhythm, quality, and deficiency of harvests ordered all material life."[5] Through climate history, economic history, or at least agrarian history, is reduced to being "one damn thing after another."

There can be little doubt that harvest yields reacted more sensitively to weather fluctuations before the advent of hybrid seed, artificial fertilizer, mechanized harvesting, and other agronomic advances of the last century. But, adverse weather conditions for one crop are not adverse for all crops, nor for the same crop on different soils. Moreover, it is not impossible that the same industrial technology that has reduced harvest variance has increased the vulnerability of other economic sectors to climatic fluctuations.[6]

The banalization of agrarian history fostered by these generalizations may still serve to emphasize an important truth when applied to a closed, technologically primitive, subsistence economy. But they cannot serve in a study of early modern Europe. Here, in most areas, the level of economic integration was sufficient—including trade, markets, inventory formation, and even futures trading—to loosen greatly the asserted links between weather and harvests and between harvests and economic life more generally. The problem, then, is to measure the actual strength and ascertain the limits of the linkages rather than to assume their importance as self-evident. Perhaps the best evidence now available to cast doubt on the "self-evident" role of short-term weather fluctuations is the meager results of efforts to link gross meteorological values to economic and demographic events.

The most thorough, systematic, and interesting studies of the human consequences of climatic fluctuations have been detailed examinations of brief periods of climatic crisis. Post's *Last Great Subsistence Crisis* and Pfister's *Agrarkonjunktur und Witterungsverlauf* are both of this kind. Post examines the post-Napoleonic war crisis in the entire Western world whereas Pfister, in the course of studying the second half of the eighteenth century,

5 Fernand Braudel (trans. Siân Reynolds), *Capitalism and Material Life, 1400–1800* (New York, 1973; orig. French ed., 1967), 18.

6 For a useful summary of adverse and favorable weather profiles see: B. H. Slicher van Bath, "Agriculture and the Vital Revolution," in Edwin E. Rich and Charles H. Wilson (eds.), *Cambridge Economic History of Europe* (Cambridge, 1977), V, 57–65.

focuses much of his attention on the crisis of 1768–1771 in the Bern region and compares it to other periods of crisis. Both analyses, although impressive in detail, are essentially descriptive when it comes to handling the human impact of climate. Having identified periods of extreme climatic adversity, they set out to uncover all of the misfortunes that befell the contemporaries to the crisis. These misfortunes are, thus, temporally associated with the climatic crisis, but they are, as the authors acknowledge, associated with much else as well. Both authors identify other factors, but in the absence of a testable model, the relative importance of the various factors remains unknown.[7]

Both Post and Pfister conclude that climate history is an important (and neglected) historical force. But by focusing attention primarily on highly unusual crises (the aftermath of the worst volcanic eruptions in centuries and a lethal combination of adverse seasons in a mountainous, marginal farming region), and by offering no means of quantifying what portion of the contemporaneous misfortunes can reasonably be attributed to climate, these authors are not really in a strong position to make such claims. Unless these crises can be shown to be something other than unique, exogenous shocks, a skeptic might feel justified in concluding that short-term climatic crises stand in relation to economic history as bank robberies to the history of banking.

That analogy expresses, of course, an extreme view, and I do not mean to trivialize the rich documentation and careful syntheses of diverse events that are found in the works of Post and Pfister. Yet both studies leave as open questions the societal significance of short-term climatic fluctuations on a year-in, year-out basis.

An alternative to the case study approach is to apply time-series analysis to the study of meteorological, economic, and other societal variables. Through regression analysis, supplemented by comparisons of means, we can theoretically identify the quantitative dimensions of societal responses to climatic fluctuations. However, the proviso "theoretical" in the preceding sentence demands emphasis, for reasons that should become clear in a

7 John D. Post, *The Last Great Subsistence Crisis in the Western World* (Baltimore, 1977); Pfister, *Agrarkonjunktur und Witterungsverlauf im westlichen Schweizer Mittelland, 1755–1797* (Bern, 1975).

review of my efforts to make such measurements for seventeenth- and eighteenth-century Dutch society.

I had constructed a series of estimated average winter temperatures for the Netherlands, extending back to 1634 the existing series of instrumental measurements, which begins in 1735. I then employed regression analysis to identify the impact of annual fluctuations in average winter temperature in the period 1634–1839 on several variables, listed in Table 1, for which long time series were available.

The first attempt took the form of a correlation of the annual fluctuation in winter temperature (measured by first differences) and the annual fluctuations in the economic and demographic variables during the years that followed. The results, shown in Table 1, gave no grounds for encouragement.

The coefficients were all close to 0, and only two, for total grain shipments from the Baltic and for rye prices, were significant at the 5 percent level. I had no reason to expect that the correlation coefficients should be high, since it was obvious that many other variables, indeed, other climatic variables, were certain to have influenced the dependent variables tested.

Moreover, this test assumed that a linear relationship existed between winter temperature and grain harvests (which we are measuring indirectly through the grain shipment and price data). Neither common sense nor the observations of agronomists offer much support for this assumption. First, it may be that the winter temperature fluctuations within a broad range around the mean were of no significance to grain production or any of the other variables. Only extreme values at either end of the temperature spectrum may have significantly affected the variables. Second, the impact of temperature changes on the variables may not have been in the same direction in these extreme zones. We know, for instance, that grain cultivation in areas that customarily experience winter frost is not adversely affected by cold winters unless the winter becomes very cold, driving the frost layer deep into the ground. However, extremely mild winters can also be unkind, since winter crops then have no frost layer to protect them from occasional cold days.

In order to test the influence of extreme winter temperatures on the economic variables I defined "extreme" temperatures as those in excess of one standard deviation from the mean. This yielded, for the period 1634–1839, seventy-three years of extreme

Table 1 Correlation Analysis of Winter Temperature and Selected Economic Variables

(X) = FIRST DIFFERENCE OF AVERAGE WINTER TEMPERATURE (Y) = FIRST DIFFERENCE OF:	R	STUDENTS' T	SIGNIFICANT AT 5% LEVEL
Rotterdam burials, 1646–1819	−.09	−1.14	
Edam burials, 1651–1784	.05	0.56	
Netherlands grain shipments, 1636–1760	−.15	−1.63	
Total grain shipments, 1636–1760	−.19	−2.05	*
Butter prices at Leiden, 1635–1839	−.03	−0.34	
Rye prices at Utrecht, 1635–1839	−.14	−1.99	*

SOURCES: Price data are drawn from N. W. Posthumus, *Inquiry into the History of Prices in the Netherlands* (Leiden, 1964), II. Grain shipments are from N. E. Bang and K. Korst, *Tabeller over skibsfart og varetransport gennem Øresund, 1497–1660 and 1661–1783* (Copenhagen, 1906–1953), 7v.; only shipments passing through the Danish Sound are included here. "Netherlands shipments" refers to grain transported in Dutch bottoms. Rotterdam burials are from A. M. van der Woude and G. J. Mentink, *De demografische ontwikkeling te Rotterdam en Cool* (Rotterdam, 1965). Edam burials are from van der Woude, *Het Noorderkwartier* (Wageningen, 1972), III, 635–638.

temperature: thirty-five below −0.1° C., thirty-eight above 3.7° C. The economic data were detrended by subtracting from them the fifteen-year moving average centered on the year in question.

Separate regression analyses were made on the years of high and low temperature. The results were interesting—but not significant. Higher temperatures within both the high and low groups were associated with a tendency for rye prices to increase. Thus, the slope of the regression equations was the opposite of the regression equation for the data as a whole. None of the coefficients was significant at the 5 percent level.

It became clear that no worthwhile results were to be achieved in the absence of a well-specified model of causation, and such models required more refined meteorological data than provided by the average winter temperature series. Luckily, one further test proved possible. The original data from which I had made the estimates of winter temperature also identified those years in which frost persisted into March, and agronomists have stressed that such an event, by delaying the spring planting and the opening of pastures to livestock, was highly unfavorable to agriculture in northwestern Europe.[8]

8 Slicher van Bath, "Agriculture," 57; J. P. M. Woudenberg, "Het verband tusschen het weer en de opbrengst van wintertarwe in Nederland," *Koninklijk Nederlandsch Meteorologisch Instituut, Mededeelingen en Verhandelingen, 50 (1946).*

In the period 1658–1757, for which data on March frosts were available, I identified two groups of years: twenty years with more than two days in March during which canals were frozen and eighty years where this was not the case. Instead of asking whether progressively longer March frost spells influenced the economic variables (which would call for regression analysis), I asked: did the economic variables have significantly different mean values in the two groups of years? That is, was the observed difference between the mean values in years with March ice and in all other years too great to be attributable to chance factors? A standard error of the difference test answers this question.

Table 2 presents the results of this test. In order for the two means to be significantly different at the 1 percent level, Z, the standard normal deviate, must exceed the value of 2.58. Only butter prices were significantly affected by March frost according to this test. The absolute difference between the mean butter prices (Column 3) is 2.56 guilders per *schippond* (80 pounds). A 99 percent confidence interval spreads 2.06 guilders (standard error × 2.58) to either side of 2.56 guilders. That is, we can be 99 percent certain that the "true" value of butter prices in years of March ice lay between 4.62 and 0.50 guilders *above* the value for other years. In all other cases the confidence intervals *included* the mean value of the years without March ice. Consequently, for all tests except butter prices, the probability is unacceptably great that whatever difference existed between the means is insignificant, i.e., could have been the product of chance.

I have drawn two conclusions from this process of trial and error. One cannot hope to achieve significant results (or fairly test the impact of climate) in the absence of a detailed causal model that specifies, to use agriculture as an example, both the critical meteorological elements and the critical periods of the farming year. In addition, testing techniques must not need to assume linear relationships throughout the full range of values.[9]

My inadequate efforts can, perhaps, be judged with sympathy when they are placed beside the statistical tests commonly used by agronomists to assess the impact of climatic fluctuations on crop yields in the twentieth century. The standard technique

9 See L. P. Smith, "The Significance of Climate Variation in Britain," in UNESCO, *Changes in Climate, Proceedings of the Rome Symposium* (United Nations, 1963), 455–463.

Table 2 Standard Error of the Difference Test with March Frost as Variable (1658–1757)

VARIABLE	1. MEAN OF MARCH FROST YEARS	2. MEAN OF OTHER YEARS	3. ABSOLUTE DIFFERENCE (1–2)	4. STANDARD ERROR OF THE DIFFERENCE	5. STANDARD NORMAL DEVIATE	6. SIGNIFICANT AT 1% LEVEL (2.58 AND ABOVE)
Butter Prices	22.96	20.40	2.56	.80	3.20	*
Netherlands Sound Shipments	18058.60	19288.56	1229.96	3491.91	0.35	
Total Sound Shipments	24468.85	24760.02	291.17	4302.95	0.07	
Rye Prices	5.79	5.49	.31	.50	0.62	
Rotterdam Deaths	1811.50	1811.21	.29	82.32	0.004	
Edam Deaths	131.20	132.55	1.35	11.39	0.118	

SOURCES: See Table 1.

is multiple regression analysis, with the independent variables typically comprising a series of climatic elements plus a trend variable to cover "technological change." The weakness of such tests, not always fully appreciated, lies in the frequent assumption of linearity (sometimes enforced by the small number of available observations) and the fact that the independent variables are correlated with each other (multicolinearity). As a result, the estimated coefficients of the various climatic variables can be unstable, with the effect of wrongly identifying the critical variables and even giving the variables a wrong sign.[10]

The second conclusion is that any test of climatic influences on economic life should take the form of a more comprehensive model that simultaneously tests the other significant variables. In the case of the dependent variable of rye prices, an econometric model could be devised in which the price is the result of equalizing supply and demand. Demand is a function of price, per capita income, and inventory formation. Supply, in the Dutch context, is a function of prices in several preceding years, imports, non-economic variables that might affect imports such as Baltic wars and blockades, and, finally, the climatic variables deemed most important to the success of the harvest. The reason such a model was not developed is obvious: data for several of the variables were not available. But, the elaboration of such a model serves the useful purpose of indicating the special conditions that are probably needed for a climatic fluctuation to dominate the short-term movement of an economic variable, such as the price of a bread grain.

In the absence of the economic data needed to perform econometric testing and in the absence of the nuanced meteorological data that could allow us to focus on a critical agronomic relationship, it is still possible to advance our knowledge about the types of influences climatic fluctuations can exert on preindustrial economies. The key is to be specific about the hypothesized causal link. If that can be achieved, simple statistical procedure can identify the likely orders of magnitude of economic responses to

10 For a non-technical discussion of these weaknesses, see Richard W. Katz, "Assessing the Impact of Climatic Change on Food Production," *Climatic Change*, I (1977), 85–96; for an example of the technique typically applied, see L. M. Thompson, "Weather and Technology in the Production of Wheat in the U.S.," *Journal of Soil and Water Conservation*, XXIV (1969), 219–224.

short-term climatic fluctuations. The following examples of such tests are all drawn from the Dutch economy in the seventeenth and eighteenth centuries.

Dairy Production. The one firm and consequential result of the naive statistical tests reported above was an association between the incidence of March frost and a rise in butter prices of about 10 percent above its fifteen-year moving average. These two fluctuations can be associated as follows: March frost, by delaying the opening of pastures, precipitates fodder shortages. Other things remaining equal, farmers responded to this shortage by reducing their herd size during the spring, and this, together with lower feeding standards for the surviving cattle, reduced the output of milk and, hence, of butter and cheese. A lowered output, *ceteris paribus,* resulted in an increased price. It happens that intermediate elements of this causal chain can be tested. Thanks to a tax levied on cattle by the province of Holland the total herd size of the province can be reconstructed twice yearly, on April 1 and October 1, from 1768 to 1805. If March frost had the hypothesized effect, the difference between the herd size reported on April 1 and that reported for the preceding October 1 should measure it with reasonable accuracy. Other, longer-term factors affecting herd size would have had relatively little influence on such a measure of herd-size fluctuations.[11]

Indeed, the mean change in herd size (cattle three years old and above) in the twelve years with average March temperatures below 3°C was −1,150, or just under 1 percent of total herd size; in the twenty-four years of mild March weather the average herd size rose slightly (260 head). These results fit our hypothesis, but, given the considerable variance, particularly among the observations for years with mild March weather, the 95 percent confidence intervals around these means overlap. That is, the observed difference could be the product of change.

When we move on to the impact of cold March weather on dairy output, more clear-cut results are achieved. The annual marketed cheese volume of north Holland, the major cheese pro-

11 A. M. van der Woude, *Het Noorderkwartier* (Wageningen, 1972), III, Appendix 10B, 654–655. The series for Holland south of the IJ was used; the figure published for 1799 is in evident error, and has been changed from 146,562 to 136,562 in the calculations reported here. *Ibid.*, II, 552–572.

ducing region of the Dutch Republic, is known from 1766 to 1804. Following the same procedures as for the test of herd size, I found that cheese output in the twelve years with cold March weather fell short of the mean value for the twenty-six other years by 1.4 million pounds, or about 8 percent. In this case the standard error of the difference statistic confirms that the 95 percent confidence intervals do not overlap.[12]

The major disturbing element in the dairying industry of eighteenth-century Holland was unquestionably epizootics. In 1714–1720, 1744–1754, and 1769–1770 farmers suffered enormous losses and dairy production plunged. Next to these catastrophes the impact of weather that adversely influenced fodder supplies pales in significance, although they do account for a good deal of the modest year-to-year fluctuations observable in the cheese production series. Perhaps farm fodder shortages did not lead farmers greatly to reduce herd size because purchased fodder supplies were readily available and could be transported cheaply. But the reduced quality of the fodder does seem to have reduced milk (and butter and cheese) output.[13]

It does not follow that the farmer suffered greatly from cold March weather, for the price elasticity of demand for dairy products seems to have been approximately unitary. When output fell by about 8 percent, butter prices rose by about 10 percent. Farm revenue showed greater stability than dairy production, which in turn was buffered from climate fluctuations by markets in both cattle and fodder.

Arable Crop Yields. A critical factor influencing the level of arable crop yields in northwestern Europe was the amount of precipitation, particularly in the winter. High rainfall levels between November and March both harmed standing winter crops and delayed spring planting, and these problems were nowhere so severe as in the low-lying parts of the Netherlands. Modern agronomists' studies of the Dutch polders have established that a high winter water table reduces soil aeration, which harms the development of root structure and reduces the amount of nitrogen

12 P. N. Boekel, *De zuivelexport van Nederland tot 1813* (Utrecht, 1929), 210–211.

13 Van der Woude, II, 585–592; J. A. Faber, *Drie eeuwen Friesland* (Wageningen, 1972), I, 155–175. In the test of fluctuations in herd size, above, data for 1770 was excluded because of the dominant influence of cattle disease.

and minerals available for plant growth. For these reasons we can hypothesize that high levels of winter rainfall, and particularly the accumulation of several successive rainy winters, reduced arable crop yields, and that this consequence was most pronounced in the areas with the worst drainage characteristics.[14]

Baars, a Dutch scholar, in a detailed study of the agrarian history of a collection of south Holland polders, provided interesting material illustrating the consequences of winter rainfall. He presented uniquely detailed evidence of the physical yields of the major arable crops produced in clay soil polders. Since we have already noted how poorly market prices serve as an indicator of harvest results, Baars' yield data, covering eleven crops and extending from 1627 into the nineteenth century, were particularly welcome. Together with the monthly record of rainfall in Holland, which begins in 1715, they permit a direct test of the causal link identified above, a link which the farmers complained about frequently in the meetings of the polder authorities.[15]

The longest continuous series of tithe records (the basis of Baars' yield calculations) belonged to the Nieuw Beijerland polder. Baars created a yield index of the eleven major crops weighted by the amount of land in the polder planted to each crop. This index, converted to ten-year cumulative averages, is displayed in Figure 1. By presenting November to March rainfall, also in ten-year cumulative averages, on an inverted scale, Baars sought to show visually the existence of a strong negative correlation between yields and rainfall. The figure also shows that changes in yields lagged behind rainfall fluctuations by three or four years, suggesting that the cumulative effects of rainfall were particularly significant. The polder relied on natural drainage until 1873, when the first pumps were installed.

A regression analysis permits us to be more precise about these relationships. In an effort to overcome some of the weaknesses of this technique identified above, I designed a multiple regression for the period 1720–1820 where Y is the yield index,

14 Slicher van Bath, "Agriculture," 57; W. H. Sieben, "Invloed van de ontwateringstoestand op stikstofhuishouding en opbrengst," *Landbouwkundig Tijdschrift*, LXXVI (1964), 784–802.

15 C. Baars, *De geschiedenis van de landbouw in de Beijerlanden* (Wageningen, 1973), 145–155. Baars very generously made the raw data available to the author to enable the regression analysis that follows.

Fig. 1 Arable Crop Yields and Winter Precipitation in Two South Holland Polders.[a]

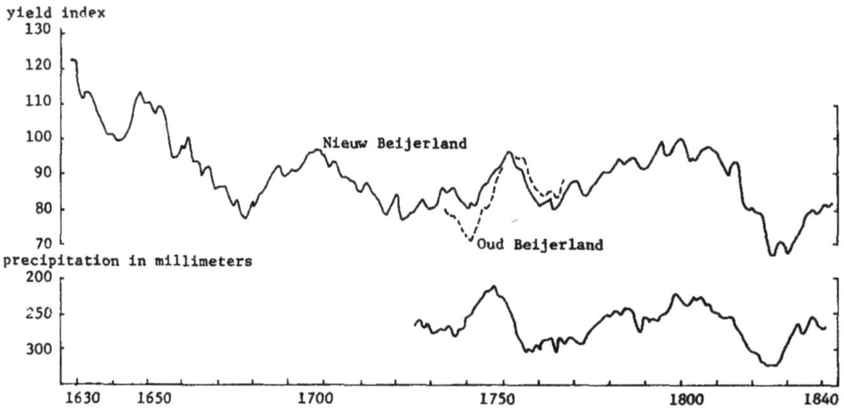

a The yield indices are weighted averages of eleven crops grown in the Nieuw Beijerland polder (1627–1841) and the Oud Beijerland polder (1723–1766). The precipitation amounts are ten-year cumulative averages of precipitation recorded from November through March at Zwanenburg, Holland.
SOURCE: Baars, *De geschiedenis van de landbouw*, Fig. 65.

influenced by a series of weather variables that were all available for the first time around 1720: X_1, average March temperature; X_2, average winter temperature; X_3, average summer temperature; X_4, winter rainfall; and X_{4c}, the cumulative rainfall of the four preceding years. Both linear and quadratic forms of the ordinary least squares regression results offered the same message: only the two rainfall variables were significant (see equation 1, t-statistics in parentheses). Equations confined to these two variables yielded nearly the same coefficients of correlation as the longer equations whereas their F-statistics showed much improvement (see equation 2).

This result is less dramatic than the visual presentation of the

1. $Y = 130.5 + .006X_1 - .343X_2 + .380X_3 - .079X_4 - .025X_{4c}$
   (4.882) (.009) (−.459) (.240) (−4.201) (−2.818)
   $R^2 = .237$ F-statistic (5,95) 5.89
   D-W = 1.4186
2. $Y = 135.88 - .081X_4 - .025X_{4c}$
   (13.583) (−4.461) (−2.869)
   $R^2 = .235$ F-statistic (2,98) 15.007
   D-W = 1.4077

data leads us to anticipate. Still, it confirms the importance of winter rainfall to the arable farming of Nieuw Beijerland; it alone explains a quarter of the annual variance in output. Moreover, the farmers of this polder, as elsewhere in the Netherlands, could not count on high grain prices to compensate them for short harvests. Local harvest conditions could influence Dutch grain markets very little, dominated as they were by international market prices. Nevertheless, these farmers were not altogether helpless in the face of the tyranny of mother nature, as the history of the adjacent Oud Beijerland polder demonstrates.

Oud Beijerland was the lowest-lying polder of the region, and the yield index (presented in Fig. 1 and beginning in 1724) reflects this fact by showing systematically lower yields than in neighboring Nieuw Beijerland until 1740. In that year the polder administration of Oud Beijerland, reacting to the deteriorating economic conditions brought on by many years of high winter rainfall, installed a windmill to pump water out of the polder and thereby lower the water table. The results are immediately apparent in Figure 1; after 1750 yields in Oud Beijerland systematically exceeded those of its neighbor, although they were by no means insulated from the influence of rainfall quantities. For this the drainage technology was not yet sufficient.

Finally, in a third polder, the Zuid Beijerland polder, the correlation between winter rainfall and yields (which, unfortunately, can be measured only in the nineteenth century) was exceedingly weak. Here, where the ground lay an average of 50 cm. higher than in Nieuw Beijerland, natural drainage was good; the farmers did not bother with the installation of pumps until 1913. Figure 2 displays the weak linkage between winter rainfall levels and crop yields.

The fascinating evidence presented by Baars demonstrates that rainy winters could affect farming very directly. However, it also demonstrates that this influence did not blanket large areas, but depended on the micro-environment, and that farmers could, and did, take steps to weaken the causal link.

Canal Transportation. Although I had little success in establishing the impact on agriculture of so crude a meteorological indicator as average winter temperature, it has proved possible to achieve more satisfactory results in the non-agricultural sector.

Fig. 2 Grain Yields in Zuid Beijerland and Precipitation from November through March.[a]

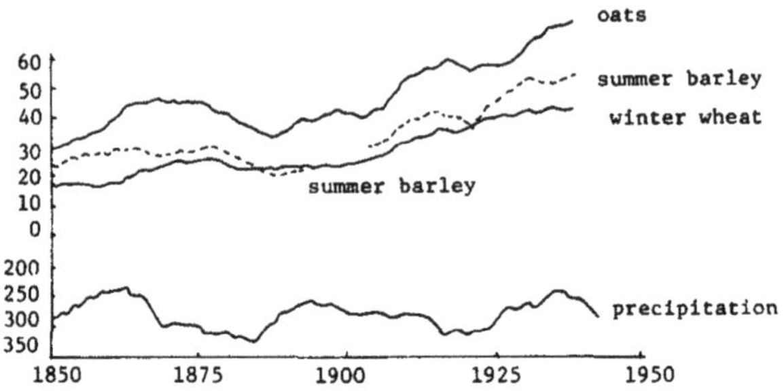

a Yields are measured in hectoliters per hectare. Precipitation, measured at Zwanenburg, is denominated in millimeters and expressed in ten-year cumulative averages.
SOURCE: Baars, *De geschiedenis van de landbouw*, Fig. 63.

Severe winters, by stopping shipping on the inland waterways that formed the communications backbone of the Dutch Republic, reduced the level of transport activity and so directly reduced the incomes of thousands of bargemen and indirectly reduced the productive capacity of the economy. My recent study of passenger transportation in the Netherlands in the seventeenth and eighteenth centuries makes it possible to be specific about the effects of winter temperature in this sector.[16]

Horse-drawn barges provided the bulk of all passenger transport in the low-lying provinces of the Netherlands, producing, at their peak in the 1660s, some 38 million passenger-kilometers of transportation annually and employing over 500 bargemen. Throughout the seventeenth and eighteenth centuries the canals connecting the major cities of Holland were closed an average of twenty-eight days per winter, but the annual totals ranged between zero and eighty-nine. A simple regression of the number of days that the canals were closed on the percentage deviation of the annual volume of passengers from its fifteen-year moving average yields the following for the period 1658–1757:

$$Y = 3.83 - 0.15X \qquad R^2 = .34$$
$$(4.72) \quad (-6.28)$$

16 De Vries, "Barges and Capitalism. Passenger Transportation in the Dutch Economy, 1632–1839," *A. A. G. Bijdragen*, XXI (1978), 35–398.

Y = Percentage deviation of passenger volume from the fifteen-year moving average

X = Number of ice-bound days

That is, a third of the annual variation in passenger travel was explained by this single variable. The coefficient of X also suggests that winter travelers, when faced by the climate-enforced suspension of services, did not delay their trips until the thaw. Trips postponed were trips cancelled.[17]

Severe winters directly affected the incomes of barge skippers in two ways. By reducing the number of passengers such winters reduced their revenues, and by indirectly increasing the price of fodder for their horses (their largest single expense) they increased their costs. The scissors movement of these two variables explained 41 percent of the annual variation of the incomes of the barge skippers between Haarlem and Leiden in the period 1780–1818. The variability in the barge skippers' incomes was substantial. Between 1780 and 1818 annual incomes varied from 350 to over 1,000 guilders, and in seven years after 1790 differed by over 200 guilders from the preceding year.[18]

Comprehensive information is not yet available concerning freight transportation. However, I have compiled the monthly toll receipts of one important artery (the sluices at Vreeswijk, passed by 2,000 to 3,000 inland vessels annually) for the period 1769–1790. They seem to confirm my *a priori* view that variations in winter temperature should have affected freight transportation less severely than passenger transportation. Unlike services, most goods can be inventoried; consequently, the reduced activity of a severe winter can be made good (at a cost, to be sure) in the non-winter months.[19]

A regression and correlation analysis shows that the average winter temperature explained only 8 percent of the variation in annual toll receipts. But this weak result is the product of two contradictory influences: severe winters sharply reduced winter shipping and explained 25 percent of the annual variance in winter receipts, whereas the severity of the winter was negatively cor-

17 *Ibid.*, 318–319.

18 *Ibid.*, 178–181, 322–323.

19 For a discussion of this issue see: Robert W. Fogel, *Railroads and American Economic Growth: Essays in Econometric History* (Baltimore, 1964).

related with the receipts of the following March–November period. Surprisingly, the net effect of a severe winter seems to have been to increase the volume of goods carried annually on the canals and rivers. (For a possible explanation, see the following section.)[20]

The short-run impact of cold winters on the transportation industry says nothing about a possible long-run impact. To anticipate a later section of this article, that the winters of the first half of the eighteenth century were substantially warmer than those of the seventeenth century implies nothing about the fortunes of barge skippers. In fact, the month-by-month record of barge passengers shows a persistent, mounting avoidance of travel in the winter months. Whereas in the mid-seventeenth century 14 percent of all trips were made in the three winter months, only 10 to 11 percent were made in that period a century later. This trend persisted into the 1840s.[21]

In the absence of direct evidence of winter temperature, a climate historian seeking proxies for winter temperature trends might be tempted to use this as evidence of the increased severity of eighteenth-century winters. Just the opposite was true. The largest number of winter travelers are observed during the decades of the severest winters. The long-term trend has little to do with climate change and much to do with the level of utilization of the economy's capacity to produce goods and services.

Fuel Prices. Today, fluctuations in average winter temperature, through their impact on the amount of heating fuel used, can play a substantial role in determining the balance of payments position of oil-importing nations. It is of obvious interest to examine the energy vulnerability of an urbanized, in-

20 The regression equations are as follows: (X = average winter temperature in °C; N = 21)

Y = annual toll receipts (December–November)
$Y = 10181 - 201X \qquad R^2 = .08$

Y = winter receipts (Dec. of preceding year, Jan., and Feb.)
$Y = 837 + 102X \qquad R^2 = .25$

Y = non-winter receipts (March–November)
$Y = 9342 - 302X \qquad R^2 = .24.$

21 De Vries, *Barges*, 337–339.

dustrial society of the *ancien régime*. Did the frequent severe winters of the little ice age strain heating fuel supplies and send prices skyrocketing?

In the pre-industrial Netherlands, the principal fuel for heating, cooking, and industrial purposes was peat. Some peat bogs lay literally at the doorsteps of the major cities, but 100 to 200 km. separated the major fields in the northern provinces from the principal markets. Peat digging was confined to a season beginning in April and ending in July, in order to allow sufficient time for the peat to dry in the sun before use. Under these conditions a wet summer could adversely affect the output. Given these characteristics of the industry, our attention is focused on what happened during an especially severe winter. No peat could be dug until the following spring and, assuming the canals were frozen, inventories that were not already near the centers of demand could not be transported.[22]

A detailed peat price series with at least monthly quotations is needed to examine the actual impact of winter temperature fluctuations. Unfortunately, the three available peat price series for the seventeenth and eighteenth centuries offer only annual averages and, since they are drawn from the records of institutional buyers, the suspicion that they reflect the existence of long-term supply contracts cannot be suppressed. Because of their quality, no formal statistical tests seem warranted, but they show the following: sharp increases in the price of peat following the cold winters of 1672, 1679, 1692, 1740, 1763, 1784, and 1795, but no significant increase following as many other cold winters, such as 1681, 1691, 1695, 1697, 1716, 1746, 1757, 1760, and 1789. Mild winters showed no tendency to depress prices.[23]

If the price series on which these observations are based can bear any analysis at all, they direct our attention to the facilities for the storage of peat and the decision-making process that determined the level of inventories—matters about which the literature is silent. One is tempted to speculate that a storage system

22 P. van Schaik, "De economische betekenis van de turfwinning in Nederland," *Economisch-Historisch Jaarboek*, XXXII (1969), 141–205; J. W. de Zeeuw, "Peat and the Dutch Golden Age. The Historical Meaning of Energy Attainability," *A. A. G. Bijdragen*, XXI (1978), 3–31.

23 N. W. Posthumus, *Inquiry into the History of Prices in the Netherlands* (Leiden, 1964), II, series 125, 142, 266.

existed that proved capable of routinely maintaining—and financing—ample inventories (witness the price stability in most severe winters of the 1690s), but that after a long string of mild winters, such as preceded the winter of 1740, the inventory level was allowed to be reduced and the system was caught "off-guard."

If adequate winter inventories could usually weaken the price-increasing pressure of severe winters, the chief impact of such winters may have been to stimulate employment in the peat bogs during the following digging season (and to increase the volume of shipping on inland waterways). Comprehensive data of the volume of annual peat output do not exist, but the two indicators of peat production that do exist (based on taxes levied on peat shipments in the provinces of Groningen and Overijssel) do not offer unequivocal support to this view. Obviously, the aggregate demand for peat did not emanate only from households; its industrial use was also great, and production trends in many Dutch industries may have influenced peat output levels far more than fluctuations in winter temperature.[24]

These case studies of the impact of short-term climatic fluctuations upon a preindustrial society do more than show the obvious. Their simple measurement techniques notwithstanding, they provide orders of magnitude for the consequences of climatic fluctuation and help identify the adjustment mechanisms—the buffers—that softened the first-approximation causal links between climate and the economy. In short, they contribute to the composition of a nuanced picture of how climatic fluctuations worked their way through a preindustrial society.

CYCLES The number of periodic climatic cycles identified by historians and climatologists are as numerous as the number of different periodicities that economists have claimed to see in long-term price series. Apart from very short oscillations, there are the well-known eleven-year cycles associated with sun spots, twenty-two-year cycles (give-or-take eleven years) put forward by Manley on the basis of English temperature data, varying periodicities based on dendroclimatological and vendange data, and longer cycles of perhaps eighty years based on long-term solar change.

24 Slicher van Bath, *Een samenleving onder spanning* (Assen, 1957), 216; W. J. Formsma et al., *Historie van Groningen* (Groningen, 1976), 326.

The present writer has not proven immune from the virus, and has put forward an extravagant claim for sixty- to one-hundred-year cycles of alternating "continental" and "maritime" climate in northwestern Europe.[25]

This is not the place for a critical examination of what appears to be, at best, a poorly integrated literature, and at worst, eloquent testimony that finding patterns in random fluctuations is comparable to a Rorschach test, and says more about the investigator than about the data. Our approach here is to ask how we can identify the importance of climatic cycles—if there are any—as a generator of business cycles, or of conjunctural movements in economic life?

A long tradition among economists, stretching from Thomas Malthus through W. Stanley Jevons, Lord Beveridge, and Eli Heckscher assumes that climate cycles are communicated to economic life via the harvests and the latter's influence on price levels. The theory of the preindustrial conjuncture associated with Labrousse regards the harvest as the critical determinant via its influence on grain prices and rural employment levels, which, in turn, influence urban incomes. If the pattern of successive harvests departs from random fluctuations around a mean, this presumably is the result of cycles inherent in the underlying climatic determinants of the harvest yields.[26]

The problem with this line of reasoning is that it has enough undeniable validity to survive in the face of test after test that shows the couplers of this train of causation to have an extraordinary amount of slack. We have already presented information to cast doubt on the existence of systematic, linear relationships between the values of meteorological phenomena and harvest size over large areas. The link between harvest size and price level is even weaker.

25 See, for an example, C. J. E. Schuurmans, "Influence of Solar Activity on Winter Temperatures: New Climatological Evidence," *Climatic Change*, I (1978), 231–237. Manley, "Temperature Trends in England, 1698–1957," *Archiv für Meteorologie Geophysik und Bioklimat*, IX (1959), 413–433; Le Roy Ladurie, "History and Climate," 146–147, 154–155; John A. Eddy, "Climate and the Changing Sun," *Climatic Change*, I (1977), 173–190. A study of ice borings also identifies cycles of nearly eighty years (not to mention periodicities of 181, 400, and 2,400 years): S. J. Johnson, W. Dansgaard, and H. B. Claussen, "Climatic Oscillations 1200–2000 A.D.," *Nature*, CCXXVII (1970), 482–483. De Vries, "Histoire du climat," 211–215.

26 Ernest Labrousse, *La crise de l'economie française à la fin de l'Ancien Régime et au début de la Révolution française* (Paris, 1944), 172–180.

Pfister's careful correlations of tithe yields and wheat prices in the Bern region between 1755 and 1797 show that the output level can explain no more than 25 percent of the variation in prices. In areas more exposed to the international market, of which the Netherlands form an admittedly extreme example, the correlations are weaker still. The tithe records and direct output records for small areas of Limburg, when correlated with local grain prices, yield coefficients of determination (R^2) no higher than .06.[27]

A more sophisticated technique for the identification of relationships between two time series is cross-spectral analysis. Lee applied this technique to central English mean temperatures in January and July and estimates of both the annual number of deaths in England and of the real wage (a sensitive reflection of the price level). None of the various indicators of cyclical correlation, such as phase angle or coherence, permitted rejection of the null hypothesis that the series were totally independent of each other.[28]

Although Lee's technique is sophisticated, the implicit model of causation being tested is not, and one can dismiss all of these tests as inconclusive because they utilize insufficiently refined data. But one basic fact of *ancien régime* economic life remains to confound those who wish to advance climatic cycles as an important generator of business cycles. International markets and, hence, international prices existed for many commodities, particularly the bread grains. Although the correlation of prices among regional markets was not always complete, neither was it often insignificant.[29]

Slicher van Bath relates how Heckscher once analyzed the relationship of Swedish harvest results and demographic variables for the century after 1680. He assumed that Stockholm grain prices could serve as a reasonable indicator of the domestic harvest. But, in that period Sweden regularly received a small portion of its grain from the Baltic grain exporters. Baltic supplies also played an important role in the formation of the international

27 Pfister, *Agrarkonjunktur*, 159. Myron Gutmann, *War and Rural Life in the Early Modern Low Countries* (Princeton, 1980), Appendix D.

28 Ronald Lee, "Model of Pre-industrial Population Dynamics with Application to England," unpub. paper (University of Michigan, 1972).

29 Walter Achilles, "Getreidepreise und Getreidehandelsbeziehungen europäischer Räume im 16. und 17. Jahrhundert," *Zeitschrift für Agrargeschichte und Agrarsoziologie*, VII (1959), 32–55.

price on the Amsterdam market. As a consequence, Swedish prices proved to be highly correlated to those at Amsterdam. The weather certainly influenced Swedish harvests, and the state of those harvests undeniably affected the broader Swedish economy. But, many other factors also affected Swedish harvests, price levels, and incomes. Even in the overwhelmingly agrarian economy of early modern Sweden no simple harvest model can account for the fluctuations of grain prices.[30]

In their studies of climatic crises, both Post and Pfister offer conclusions advancing climatic fluctuations as a generator of economic cycles. Pfister writes: "The underlying factor of climate appears to have influenced the economic conjuncture and the population movement of earlier centuries in a much stronger measure than has hitherto been accepted." Post advises that: "For the economic historian, the inability to explain anomalous weather patterns with quantitative exactness is less critical than it is to remember that the elements can become an uncontrolled independent variable affecting the intensity, duration, and perhaps the occurrence of trade cycles." The crises that they bring to our attention as having such consequences are undeniably dramatic, but if they are to be put forward as generators of economic cycles it is not clear to me how this can be in any other context than the theory of economic cycles associated with the work of Slutsky or expressed in the Klein-Goldberger model. Both see them as a statistical artifact created by the cumulative effects of random shocks to an economy.[31]

CLIMATE CHANGE The possibility that long-term changes in climate might play a role in historical explanation constitutes the most exciting dimension of climate history. Correspondingly, the fragments of available evidence concerning such long-term changes have motivated several historians to bold speculations.

The fourth century A.D. encompassed the coincident collapse of centralized empires from Rome, across Eurasia, to China. Or, were these events connected? Lopez, in *The Birth of Europe*, rumi-

30 Slicher van Bath, "Agriculture," 64; Eli F. Heckscher (trans. Göran Ohlin), *An Economic History of Sweden* (Cambridge, Mass., 1954), 133–140.

31 Pfister, *Agrarkonjunktur*, 190; Post, *Crisis*, 26. Among the works seeking to demonstrate the ability of random shocks to generate cyclical movements are: Eugen Slutsky, "The Summation of Random Causes as the Source of Cyclic Processes," *Econometrica*, V (1937), 105–146; Irma Adelmann and Frank Adelmann, "The Dynamic Properties of the Klein-Goldberger Model," *Econometrica*, XXVII (1959), 596–625.

nates: "It may well be that the study of climate could help us to understand the apparent coincidence of the main long-run demographic and economic fluctuations throughout Eurasia."[32]

The fourteenth century witnessed a combination of demographic, economic, medical, and political disasters that brought a brilliant medieval civilization to its knees. Lopez, who began his study of the birth of Europe with the fourth-century disorders just mentioned, closes it with a description of numerous climate-induced setbacks, and pronounces the fourteenth century as "The crest of a [climatic] pulsation."[33]

At the end of the sixteenth century the curtain was drawn on the Mediterranean basin's long reign as Europe's economic and cultural center of gravity. Braudel, whose classic *The Mediterranean and the Mediterranean World in the Age of Philip II* is dedicated to coming to terms with this seminal turning point, dared to take a position "at the utmost limits of prudence" when he wrote in the first, 1949, edition that "the roots of the social crisis caused by the food shortage that dominated the end of the century may have lain in an alternation, even a very slight one, in the atmospheric conditions." Twenty years later, the limits of prudence had receded a great distance; Braudel could pronounce that "The 'early' sixteenth century was everywhere favored by the climate; the latter part everywhere suffered atmospheric disturbances."[34]

In *Capitalism and Material Life,* in which Braudel surveys elements of world history, he is fascinated, much as was Lopez, by "the possibility of a certain physical and biological history common to all mankind [which] would give the globe its first unity well before the great discoveries, the industrial revolution, or the interpenetration of economies."[35]

The cessation of economic expansion and the social and political instability common to many European states in the first half of the seventeenth century has inspired several historians to formulate concepts of "general crisis." Not two years after the first such attempt Utterström marshalled a wide variety of evidence to argue that this was no Malthusian crisis of population

32 Robert Lopez, *The Birth of Europe* (New York, 1966; orig. French ed., 1962), 29.

33 *Ibid.*, 396.

34 Braudel (trans. Siân Reynolds), *The Mediterranean and the Mediterranean World in the Age of Philip II* (New York, 1972; orig. French ed., 1949; revised ed. 1966), 270, 275.

35 Braudel, *Capitalism,* 19.

growth outstripping resources, but a crisis of climate change, which, as it were, shifted the supply curve violently to the left. More generally, it is claimed, as Parker does in *Europe in Crisis,* that the culmination of the little ice age made the seventeenth century something it otherwise would not have been.[36]

Davis, in his survey, *The Rise of the Atlantic Economies,* chooses his words carefully when discussing climatic influences in the text, but in his introduction he proclaims, with regard to the early modern period, that "the changing climate . . . was a factor of the utmost economic importance which is only now beginning to be cautiously approached by historians."[37]

Finally, Europe in the aftermath of the Napoleonic Wars was plunged into a sharp depression, and although this depression was brief, the next three decades have appeared to many historians as gloomy, socially tense times—phenomena commonly explained as the growing pains of early industrialism. But the lure of climate history reaches even here. Post poses the question:

> To what extent were the political and social troubles which peaked during the period 1800–50 prompted by the economic consequences of harsher climate, as well as by the discontinuities associated with early industrialization?

He focuses his attention only on the second decade of this period, but he is confident that the crisis "owed its supranational scope only to a minor extent to the newer economic forces, i.e., industrialization . . ." and much more to factors that "were meteorological in nature."[38]

Lopez, Braudel, and Davis, not themselves historians of climate, and able to offer only illustrative evidence, urge their readers to be receptive to an anticipated historiographical breakthrough that will give climate change a more central position in historical explanation. But these expectations are not universally shared. Postan, a medievalist, persistently subordinated climatic to economic explanations of the late medieval crisis whereas Slicher van Bath, an agrarian historian, remains unconvinced that

36 Utterström, "Climatic Fluctuations," 39–44; Geoffrey Parker, *Europe in Crisis, 1598–1648* (London, 1979), ch. 1.

37 Ralph Davis, *The Rise of the Atlantic Economies* (Ithaca, 1973), xii.

38 Post, *Crisis,* xii, xi.

climate change played an important role in the seventeenth century decline of yields which he has done so much to document. Most influential has been the judgment of Le Roy Ladurie that "in the long term the human consequences of climate seem to be slight, perhaps negligible, and certainly difficult to detect."[39]

Since few economic historians care to challenge the historical existence of climatic change, *per se,* the fate of climate change as a significant variable in historical studies hinges on the successful development of a means of measuring its influence. It might be fair to say that historians are psychologically ready, even eager, for the rise of climatic change as a vehicle of long-term historical explanation, but do not possess the means of distinguishing its impact from among the many other variables at work on human society.

In theory, the historical impact of climatic change is measurable, just as are shorter fluctuations. One can imagine a long time-series analysis in which, say, the little ice age is a dummy variable. In broad terms, the task of such an analysis would be to show that, given appropriate confidence intervals, the mean values of little ice age and non-little ice age observations differ. Alternatively, we could imagine an exercise of "counterfactual" history in which a non-existent climate is substituted for that which in fact occurred, and the difference in estimated crop yields and population growth is measured.

In practice, there is little hope of performing such analyses. The data do not suffice for such approaches, probably not even in the twentieth century. But, the problem is not simply one of data; a more fundamental weakness of such approaches is the static concept of human society that they reflect. For example, the increased rainfall of the fourteenth century or the colder

39 M. M. Postan, "Die wirtschaftlichen Grundlagen der mittelalterlichen Gesellschaft," *Jahrbücher für Nationalökonomie und Statistik*, CLXVI (1954), 180–205; Slicher van Bath, *The Agrarian History of Western Europe, A.D. 500–1800* (London, 1963), 7, 161; *idem,* "Oogsten, klimaat en conjunctuur in het verleden," *A. A. G. Bijdragen,* XV (1970), 118–133; *idem,* "Agriculture," 57–65. Le Roy Ladurie (trans. Barbara Bray) *Times of Feast, Times of Famine: A History of Climate Since the Year 1000* (Garden City, 1971), 119. Proponents of the climatic variable in history bridle at Le Roy Ladurie's skeptical approach. See Post, "Meteorological Historiography," *Journal of Interdisciplinary History,* III (1973), 721–732; Robert H. Claxton and Alan D. Hecht, "Climatic and Human History in Europe and Latin America: An Opportunity for Comparative Study," *Climatic Change,* I (1978), 195–203.

weather of the seventeenth century are presumed to have reduced yields and increased mortality by some amount when compared to other periods. Yet, since even the effects of annual climatic fluctuations have proved so difficult to measure with accuracy, this approach hardly seems promising when applied to longer periods in which many parameters are shifting.

As an alternative approach I suggest, first, that the human consequences of climatic change might better be studied by borrowing a concept developed in recent years by economic historians of technological change. The concept of "learning by doing" emphasizes "the possibility of human response to increasing scarcities—responses in which adaptations and modification of behavior, substitution of abundant for scarce materials, and even learning play an important role."[40]

Applied to the study of the impact of climate change, this concept directs attention, taking agriculture as an example, away from measuring the change in the average yield of wheat toward identifying alterations in the crop mix. Indeed, long-term change in European agriculture is at least as much a story of the introduction of new crops and changes in the relative importance of crops as it is a story of increasing yields in existing crops.

For example, in the Netherlands, buckwheat, a hardy crop which requires only a short growing season, is thought to have been of little importance before the mid-sixteenth century, is known to have spread rapidly in the following century (the build-up of the little ice age), and to have declined in importance in the eighteenth century (when winters were strikingly milder). In England, a detailed study of agrarian change in a Midland district documents a tendency toward crop diversification between the mid-sixteenth and mid-seventeenth centuries. Farmers devoted progressively more acreage to spring crops and relatively less to the winter crops which had predominated at the beginning of the period. These substitutions of barley and oats for wheat and, to a lesser extent, rye are explained as a response to the shift in demand toward the cheaper grains brought on by the reduced purchasing power of the population.[41]

40 Nathan Rosenberg, *Perspectives on Technology* (Cambridge, 1976), 281.
41 Victor Skipp, *Crisis and Development. An Ecological Case Study of the Forest of Arden, 1570–1674* (Cambridge, 1978), 44–49.

I have no reason to question this interpretation, but it is also possible that farmers made these adjustments in part to reduce their exposure to a climate that was more frequently threatening to winter crops. If this could be sustained, it would provide an example of a response to an increasingly adverse climate that succeeded in providing a more stable food supply than had existed during the climatically more benign "late medieval optimum."

The second suggestion to better identify the impact of long-term climate change concerns the excessive emphasis placed by most historians of climate on identifying changes in meteorological magnitudes—how much colder, or wetter—and trying to locate the consequences of such changes. This approach fails to appreciate that the consequences of climate changes do not flow only, probably not even primarily, from differences in *level*; they also flow from differences in *variance*. Every economic system must accommodate some measure of temperature and rainfall fluctuation, and the 1°C difference between cold and warm centuries, which is as much as studies of the last millennium have been able to verify, can be accommodated without enormous difficulty into even primitive—perhaps especially by primitive—technologies. But economic life and particularly farming can also be approached as a problem of coping with risk. Production decisions depend on the decision-maker's (implicit) assessment of the probabilities of crop failure. The same climate change that alters the mean values can also alter the relative frequency with which extreme climatic conditions occur. The actual risk-averseness of farmers and businessmen will vary according to circumstances, but at any given level of risk-averseness we can expect that long-term changes in the variability of climatic phenomena will induce adjustments in economic behavior.

This process of adjustment can be formalized by using the model of risk minimization strategy developed by McCloskey in his study of open-field agriculture. He accounts for the existence, and persistence, of scattered plots in the open fields of medieval and early modern Europe by pointing to their role in reducing the variance of a farmer's yield and, hence, his exposure to disaster. Figure 3 describes the economic consequences of consolidated farms and farms composed of scattered holdings: scattering reduces the mean yield (the cost), but yield variability is also less, and this reduces the incidence of disastrous crop years, i.e., po-

Fig. 3 Diagrammatic Representation of the Cost and Benefit of Scattered Holdings.

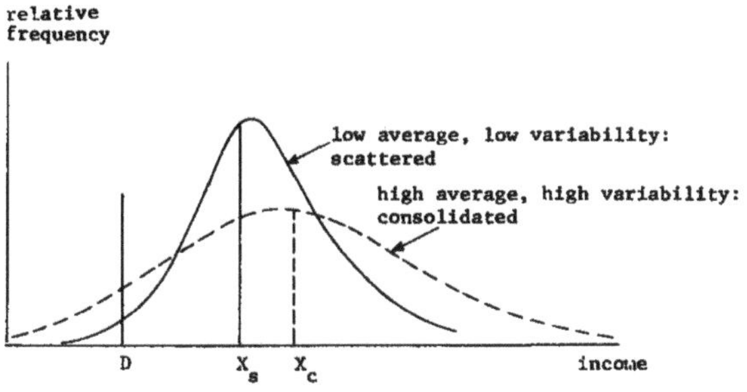

sitions to the left of line D (the benefit). McCloskey goes on to calculate the number of plots into which a farm should be divided in order to balance optimally the costs (lower average yields) and the benefits (fewer disasters).[42]

The strategy of scattering is obviously related to climate. A farmer seeks to have holdings on different soils and at different elevations so that any given climatic misfortune will not reduce the yields on all of his fields at the same time. McCloskey assumes that the underlying climatic variances faced by the farmer remain constant. But, if these variances were to change over time, McCloskey's calculation of optimal scattering would also require periodic revision.

How big have been the changes in the variability of key meteorological phenomena? A systematic answer must be left to climatologists, but one example is provided by the analysis of my time series of Dutch average winter temperature from 1634 to the present. This series was broken down into a number of cycles of alternating periods of predominantly frontal circulation, with "continental" climate, and zonal circulation, with "maritime" climate. Table 3 displays the results. The mean winter temperatures differ by about 1°C; the summer temperatures (until the mid-twentieth century) by much less.

But note the systematic and statistically significant difference in the variability of the annual temperatures within each period.

42 Donald N. McCloskey, "English Open Fields as Behavior toward Risk," in Paul Uselding (ed.), *Research in Economic History*, I (1976), 124–170.

Table 3 Temperature Means and Standard Deviations
Pattern A: Continental Climate, Frontal Circulation
Pattern B: Maritime Climate, Zonal Circulation

PATTERN	YEARS	AVERAGE SUMMER TEMP.		AVERAGE WINTER TEMP.	
		MEAN	STD. DEV.	MEAN	STD. DEV.
A	1634–1698	15.91[a]	—	1.43[b]	1.99[b]
B	1699–1757	15.61	0.72	2.50	1.55
A	1758–1839	16.14	0.94	1.61	1.93
B	1840–1939	16.17	0.88	2.32	1.56
A	1940–1970	17.04	1.25	1.91	1.98

a The summer temperature data used for this period and half of the next period are estimates based on the Angot wine harvest series. They are by no means exact. See de Vries, "Histoire du climat," 211–213, for discussion.
b The mean temperature and the standard deviation are slightly underestimated because of the downward bias of the regression equation in estimating winter temperature in the very warmest years. This bias affects the following period, up to 1734, to a lesser extent.

Figure 4 displays this difference in generalized form. Observe that a shift from "maritime" to "continental" climate, which reduces the mean winter temperature by 1°C, greatly increases the incidence of severe winters. Cold average winter temperature, by itself, does not bring disaster to most farming activities, but other climatic variables must also show this sort of pattern. If so, we can indicate the risk implications by adding a vertical line (D) to Figure 4. This disaster line (placed arbitrarily in this example) identifies the extreme temperature below which harvests will fail to satisfy the minimum needs of the producers, or some other definition of disaster. The frequency with which disasters occur under the two climatic regimes can readily be determined. In effect, a shift from maritime to continental climate increases the cost of a given level of disaster avoidance (the cost of moving the disaster line from D to D'); the measures a farmer might take to achieve such a reduction of risk are a further scattering of his holdings and changes in his crop mix which would have the effect of further reducing the variability of his annual yields.[43]

A concrete example of these adjustments might be provided by the famous winter of 1740. This severe winter, although no

43 For similar evidence based on Swiss and French vendange data see: Anne-Marie Piuz, "Climat, récoltes et vie des hommes á Genève, XVIe–XVIIIe siècle," *Annales,* XXIX (1974), 608.

Fig. 4 Generalized Distribution of Average Winter Temperatures in Continental-Frontal (A) and Maritime-Zonal (B) Periods.[a]

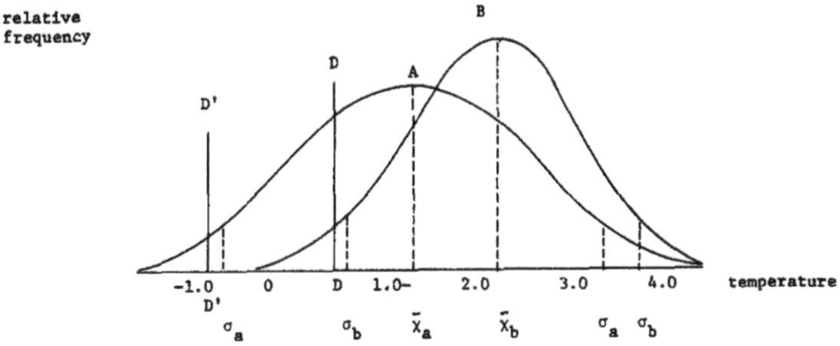

a $\bar{X}$ = mean; α = one standard deviation from the mean; D,D' = "disaster" thresholds.

colder than at least four other winters in the preceding century, has long been the focus of historical attention as a particularly disastrous climatic event. We noted earlier how peat prices skyrocketed in that year; contemporaries' assessments of the resulting crop failure have no parallel in the surrounding centuries; economic historians have regarded the ensuing crisis as instrumental in forcing the spread of potato consumption among the lower classes.

What distinguished the winter of 1740 was not so much its absolute severity as its chronological position at the end of a half century of predominantly maritime climate and, specifically, a thirty-one-year period in which even moderately severe winters rarely occurred. In those decades of reduced climatic variability did farmers develop technologies and practices dependent on "maritime level" probabilities of encountering climatic extremes? Could such an adjustment account for the fact that peat prices rose in 1740 by three times more than the largest fluctuations of the preceding twenty years, but did not jump in most of the severe winters at the end of the seventeenth century, which was also the end of a long period of "continental variability"?

This probabilistic approach focuses attention on the human context of climatic change. Such research is well suited to historians for whom man, rather than nature, is usually regarded as the proper object of study. It is also crucial to any study of the impact of climatic change, for it is undeniable that the impact of the same natural phenomenon can vary enormously depending on the characteristics of the human society that it touches.[44]

This truth is vividly illustrated in a photo given prominent attention in a recent issue of a leading Dutch newspaper. The photo shows a highway at the Dutch-Belgian border during the severe and unusually snowy winter of 1979. On the Dutch side of the frontier the road surface is clear of snow and easily driveable whereas the Belgian portion is indistinguishable from the snow-covered landscape. The Dutch article accompanying the photo went on to describe how daily life in Belgium had been brought to a standstill. It blamed this on a Belgian policy of not maintaining equipment to fight infrequent severe winters and, reflecting that policy, the exhaustion early in the winter of road salt supplies and the impossibility of getting more until the canals and roads were again usable. The neatly-cleared Dutch roads reflected a different perception of the costs and benefits of maintaining equipment and building up road salt inventories. Clearly, the paralysis of road transportation in Belgium was occasioned by the winter of 1979, but it was not an inevitable consequence of it. Is it reasonable to assume that the next severe winter, if it occurs soon, will have the same effect? After how many severe winters will the decision-makers reassess the probabilities underlying their cost-benefit analysis and take measures that will shift line D in Figure 4 to the left?[45]

The influence of climatic change is real; the difficulty in detecting it may come from the fact that we have tended to look in the wrong places. So long as the focus of attention is on measuring "the harm done," climate history must remain wedded to the study of short-term crises. The reason that the harm done, over longer time periods, is often (but not always) so slight is that in the long run societies adjust. Even primitive technologies have some capacity for adaptation. In measuring the human consequences of climate change our attention should be focused on those processes of adaptation.

44 A recent study emphasizing this point is Michelle Burge McAlpin, "Dearth, Famine, and Risk: The Changing Impact of Crop Failures in Western India, 1870–1920," *Journal of Economic History*, XXXIX (1979), 143–157.

45 *NRC-Handelsblad* (Weekeditie voor het buitenland) 30 Jan. 1979, 6.

Helmut E. Landsberg

Past Climates from Unexploited Written Sources

In many parts of the globe climatic information has been regularly published since the organization of governmental meteorological services. This organization took place in major countries in the middle or toward the end of the nineteenth century. In the United States weather observations since the establishment of a Weather Bureau (now the National Weather Service) by an act of Congress form the basis of the official climatic record. In developing countries weather services did not start until after World War II, but a number of the colonial administrations maintained weather records from the end of the nineteenth or the beginning of the twentieth century. These records may be regarded as superior material for climatological assessment because of the standardization of procedures and instruments through international agreement, first under the aegis of the International Meteorological Organization (IMO), and later by its successor, the World Meteorological Organization (WMO), one of the specialized agencies of the United Nations.[1]

If we agree that climate can be established by statistical measures from individual weather (meteorological) observations, it is clear that the information gathered during the last century is fairly homogeneous. Moreover, most of it is published, easily accessible, and, over the past few decades, has been transferred to automatic data-processing modes. It is therefore not surprising that the climatological literature, even for the intriguing studies of climatic fluctuations, has relied heavily on this kind of material.

Helmut E. Landsberg is Professor Emeritus in the Institute for Physical Science and Technology at the University of Maryland. He is the author of *Weather and Health* (New York, 1969).

Research for this study was supported by the Climate Dynamics Program (Division of Atmospheric Sciences) of the National Science Foundation under Grant No. 77-19218 A01. The author is obliged to his colleague, Stephen Brush, for some helpful suggestions. Brian Groveman did most of the work in reducing the long temperature series. Clare Villanti prepared the graphics and Katherine Mesztenyi typed the manuscript.

1 The U.S. Weather Bureau was established by Joint Resolution, 41st Congress, 2nd Sess. (16 Stat. at Large 369), Feb. 9, 1870. IMO was established in 1873. Hendrik Gerrit Cannegieter, "The History of the International Meteorological Organization 1872–1951," *Annalen der Meteorologie*, 1 (1963), 280. WMO was established by Convention, 11 Oct. 1947 at Washington, D.C.

There is, however, a wealth of additional meteorological information available for several centuries prior to these official observations. Some of this information has been resurrected and interpreted, but much remains to be exploited.[2] This article discusses this hoard of weather information, its nature and accessibility, and its quality. Admittedly, the validity of this material is mixed. Some of it is readily compatible with current observations; other data give only general hints about historical weather events. Yet the greatest limitation is geographical. The vast majority of sources are from Europe and from North America; some of them are from the Far East. This is not to say that there are no sources from other areas but, if they exist, they have yet to be discovered and evaluated.

Prior to the invention and development of meteorological instruments in the seventeenth and eighteenth centuries most of the weather information comprised two categories. One related catastrophic or damaging weather events; the other gave chronological listings of the daily weather. Most of these sources were localized. Beginning with the eighteenth century there were a number of chronicles exclusively devoted to weather events. Many of these have not been referred to in the standard works dealing with problems of climatic change. Among them is a survey by Krüger covering seasonal weather overviews for the seventeenth century with a strong regional bias for the southeastern Baltic littoral. For Bohemia (now Czecholslovakia) there are compilations by Katzerowsky and Strnadt, for Hungary by Réthly, and for Italy by Toaldo. A very useful chronicle on the ice conditions on the Baltic has been published by Betin and Preobrajenskii. One of the most useful sources of this type is a book by Pilgram, a Viennese clergyman, who attempted to assess the probability of the recurrence of various types of severe weather such as floods and droughts. He used dozens of sources, including books, monographs, and newspapers among published material, but, more remarkably, fifty chronicles and year books kept in monasteries or by chapter clergy. His compilation has been done

2 Hubert H. Lamb, *Climate: Present, Past and Future* (London, 1973), I, 612; *ibid.* (London, 1977), II, 835.

with considerable care but has a distinctly Central European, specifically Austrian, bias.[3]

Less systematic, but occasionally valuable, are reports of weather included in histories of military campaigns, which into modern times have been influenced by weather events.[4]

Some outstanding flood heights have been chiseled into stone on bridges and buildings, especially in Europe. In any climatic zone floods follow statistical rules for extreme values. This permits an assessment of whether or not historical floods were within the statistical distributions established for floods in modern times.[5]

More valuable than such sporadic information for judging climatic fluctuations are regular observations of weather on a day to day basis. The earliest existing weather diary is by Merle in Driby, Lincolnshire, England, covering the period from 1337 to 1344. A daily weather record from 1399 to 1400 probably came from northern Switzerland or the adjacent region of France. From the sixteenth century onward there are a substantial number of such diaries. The keeping of these systematic weather records was greatly facilitated by printing of so-called writing calendars (Schreibkalender) in which weather records were kept in juxta-

3 Georg(ius) Krüger, "Prodromus Aurorae Boreae sive Historiae Meteorological Teutonico-Curlandiae" (Riga, 1700); W. Katzerowsky, *Die meteorologischen Aufzeichnungen der Leitmeritzer Stadt aus den Jahren 1564 bis 1607* (Prague, 1886), 29; *idem, Die meteorologischen Aufzeichnungen des Leitmeritzer Rathsverwandten Anton Gottfried Schmidt aus den Jahren 1500 bis 1761* (Prague, 1887), 29; *idem, Meteorologische Beobachtungen aus den Archiven der Stadt Leitmeritz* (Leitmeritz, 1895), 30; Anton Strnadt, *Chronologisches Verzeichniss der Naturbegebenheiten im Königreich Böhmen (633–1700)* (Prague, 1790), 235; Antal Réthly, *Idöjaras események és elemi esapások magyarorszagon* (Budapest, 1962–1970), 2v.; Guiseppe Toaldo, *Saggio Meteorologico* [trans. as *Essai Meteorologique* by Joseph D'Aquin] (Chambery, 1784), Weather Chronicles, 233–253; Vasily V. Betin and Yu. V. Preobrajenskii, *Surovost zim v Evrope i ledovitost Baltiki* (Severe Winters in Europe and Ice Cover of the Baltic) (Leningrad, 1962), 109; Anton Pilgram, *Untersuchungen über das Wahrscheinliche der Wetterkunde durch vieljähriche Beobachtungen* (Wien, 1788), 608.

4 K. S. Douglas, Hubert H. Lamb, and C. Loader, *A Meteorological Study of July to October 1588: The Spanish Armada Storms* (Norwich, 1978); J. M. Stagg, *Forecast for "Overlord"* (London, 1971), 128; David M. Ludlum, "The Weather of American Independence," *Weatherwise*, XXVI (1973), 152–159; *ibid.*, XXVII (1974), 162–168; *ibid.*, XXVIII (1975), 118–121, 147; *ibid.*, XXVIII (1975), 172–176; *ibid.*, XXIX (1976), 236–240, 288–290; Walther Stöbe, "Forecasting for the Escape of Scharnhorst and Gneisenau," *Meteorological Magazine*, CVII (1978), 321–338.

5 Emil J. Gumbel, *Statistical Theory of Extreme Values and Some Practical Applications* (Washington, D.C., 1954), 51.

position to the printed planetary, lunar, and zodiacal constellations. They were kept so as to relate weather to celestial configurations. These early meteorological efforts were strongly influenced by astrology. Even as distinguished an astronomer as Kepler seems to have had such notions.[6]

A number of such weather diaries have been transcribed and analyzed by Lenke. They include the observations of Brahe at Hven, Treutwein at Fürstenfeld, and Fabricus in East Friesland. By far the longest and most elaborate of these non-instrumental weather records is the one kept by and on behalf of the Landgrave Hermann IV of Hesse (Uranophilus Cyriandrus) at Kassel and Rotenburg. Analysis of the information on various weather events indicates that conditions in the 1621 to 1650 interval did not differ much from modern times except that there was a higher frequency of precipitation. In North America the first systematic weather observations were made in 1644–1645 by John Campanius Holm, the chaplain of the Swedish military expedition, at old Swedes Fort (now Wilmington, Delaware). A modern evaluation of these data was made by Havens.[7]

General diaries of the seventeenth and eighteenth centuries also contain casual or even fairly regular references to weather, depending on the occupation of the writer. Comments on severe seasons or violent weather abound. Many of these sources remain unexploited. They are widely scattered in libraries, archives, and

6 William Merle, "Consideraciones temperici pro 7 annis (1337–1344)," printed as *The earliest known Journal of Weather* (reproduced and translated under the supervision of G. J. Symons), (London, 1891); E. N. Lawrence, "The Earliest Known Journal of the Weather," *Weather,* XXVII (1972), 494–501; Ralph H. Frederick, Landsberg, and Walter Lenke, "Climatological Analysis of the Basel Weather Manuscript: A Daily Weather Record from the Years 1399 to 1401," *Isis,* LVII (1966), 99–101; Fritz Klemm, "Über die Frage des Beobachtungsortes des Baseler Wettermanuskriptes von 1399–1406," *Meteorologische Rundschau,* XXII (1969), 83–85; Lynn Thorndike, "A Daily Weather Record from the Years 1399–1401," *Isis,* LVII (1966), 90–99; Christian Frisch (ed.), *Joannis Kepleri Opera Omnia* (Frankfurt, 1868), VII, 618–653.

7 Lenke, *Das Klima des 16 und Anfang des 17 Jahrhunderts nach Beobachtungen von Tycho de Brahe auf Hven, Leonard III Treutwein in Fürstenfeld und David Fabricius in Ostfriesland* (Offenbach, 1968), 49; *idem, Klimadaten von 1621–1650 nach Beobachtungen des Landgrafen Hermann IV von Hessen* (Offenbach, 1960), 31; Weather Observations published by Thomas Campanius Holm (Johann Campanius Holmensis), "Om Wäderleken och Ehrsens Tijder uti Virginien och Nya Sverige" (Stockholm, 1702). Reprinted in Gustav Hellmann, *Neudrucke von Schriften und Karten über Meteorologie und Erdmagnetismus* (Berlin, 1901), 53–58. James M. Havens, "The 'First' Systematic American Weather Observations," *Weatherwise,* VIII (1955), 116–117.

collections of historical societies. Their merit is that they cover a wide geographical spread, whereas the regular early weather observations come from single localities often far apart and at different time intervals. The casual weather references can serve as a cross check on severe weather situations and on the dating of events.

In areas where boat traffic has been traditionally important seasonal weather can often be judged by the freezing and opening dates of waterways and harbors. Examples are the Dutch canals, Hudson's Bay, and the Hudson River. Often the length of the frozen condition can, by comparison with modern observations, be related to the prevailing winter temperatures. Systematic ice records over several centuries for Lake Suwa in Japan, compiled by Arakawa, have been used to estimate Tokyo's winter temperatures. In evaluating such observations the excellent rules of Moodie and Catchpole should be heeded.[8]

A particularly useful source, often meticulously kept, is the log book of a vessel. Some of the material in naval archives has been exploited for climatological information since the last century. Especially in the era of the sailship information was usually recorded about wind force and direction as well as about barometric pressure. These records have helped to trace the climatic variations over the North Atlantic in the last two centuries.[9]

Although many studies of climatic fluctuations have used crop records as criteria of summer weather conditions, except for cases of severe drought, I have considerable doubts about this kind of source, as I discuss below.

Far more objective than the best descriptions of weather are instrumental records. A number have survived even from the second half of the seventeenth century. The record which is most compatible with recent observations is that reconstructed by Man-

8 Hidetoshi Arakawa, "Fujiwhara on Five Centuries of Lake Suwa in Central Japan," *Archiv für Meteorologie Geophysik and Bioklimatologie*, VI (1954), 152–166. Tables of closing and opening of the Hudson River may be found in: Joe Munsell, *Annals of Albany* (Albany, 1850), I, 326–327; *idem.*, *Collections on the History of Albany* (Albany, 1871), IV, 78; H. M. van den Loon, H. J. Krijnen, and C. J. E. Schuumans, "Average winter temperatures at DeBilt (The Netherlands): 1634–1977," *Climatic Change*, I (1978), 319–330; D. W. Moodie and A. J. W. Catchpole, "Environmental Data from Historical Documents by Content Analysis," *Manitoba Geographical Studies*, V (1975), 119.

9 Matthew Fontaine Maury, *Physical Geography of the Sea* (New York, 1855), 389; Lamb, *Climate*, II, 25–26.

ley for central England. There are also interpretable readings for Paris, Tübingen, Ulm, and Kiel for that century.[10]

In the early eighteenth century meteorological observations increased in number in Europe. We have observations from Berlin from 1700 and DeBilt from 1706 onward. With later interruptions data from Ulm begin in 1710, from Copenhagen in 1751, and Stockholm in 1756. In North America the earliest barometric records date from 1725 to 1726. The first thermometer readings come from Philadelphia in 1731, followed by data from Charleston, South Carolina in 1738, Cambridge, Massachusetts in 1742, and Nottingham, Maryland in 1753.[11]

Two very severe winters in Europe gave considerable impetus both to the development of meteorological instruments and to the zest for systematic weather observations. The extremely cold winter of 1708–09 stimulated some of the work by Gabriel Fahrenheit, who made the first reliable thermometers.[12]

Early information on precipitation is almost exclusively restricted to extremes, and references to excessive rain or drought abound. Although it is known that rain gauges were used in India in the fourth century B.C. and in classical antiquity in the Mediterranean countries, no records have survived. In China general descriptions of stream flow have been kept for centuries and have been used for the general classification of precipitation conditions. In the fifteenth century rain gauges were known to exist in Korea, but recorded information did not begin until 1770, when the long observational record for Seoul started. In Europe the earliest in-

10 Gordon Manley, "Central England Temperatures: Monthly Means 1659–1973," *Quarterly Journal of the Royal Meteorological Society*, C (1974), 389–405; Jacques Dettwiller, "L'évolution séculaire de la température à Paris," *La Météorologie*, XIII (1978), 95–130; Lenke, *Bestimmung der alten Temperaturwerte von Tübingen und Ulm mit Hilfe von Häufigkeitsverteilunge* (Offenbach, 1961), 71; Samuel Reiher, "Observationes tricennales circa frigus hymeale ex Ephemeridibus," *Miscelleni Berolinensia* (Berlin, 1770), 379–384.

11 A. Labrijn, "Het klimaat van Nederland gedurende de laatse twee een halve eeur," *Koninklijk Nederlandsch Meteorologisch Instituut Mededelingen en Verhandlingen*, CII (1945); Peter Horrebow, *Tractatus Historico-Meteorologicus* (Copenhagen, 1780), 364; H. E. Hamberg, "Moyennes mensuelles et annuelles de la température et extrêmes de températures mensuels pendant les 150 annees 1756-1903 à l'observatoire de Stockholm," *Kunglis Svenska Vetenskapakademiens handlingar*, XL (1906), 1-59; H. E. Landsberg, C. S. Yu, and L. Huang, *Preliminary Reconstruction of a Long Time Series of Climatic Data for the Eastern United States* (College Park, 1968), 30.

12 Georg Wolffgang Kraft, *Wahrhaffte und Umständliche Beschreibung und Abbildung des im Monath Januarius 1740 in St. Petersburg aufgerichteten merkwürdigen Hauses von Eiss* (St. Petersburg, 1741), 30.

terpretable precipitation records started in 1708 at Zürich; another series began in England in 1725. In North America the earliest precipitation measurements date from 1738.[13]

In 1723 Jacob Jurin, the secretary of the Royal Society of London, published a scheme for making meteorological observations and tried to enlist the scientists of the day to send their observations to the Royal Society. However, his appeal evoked little response. The time was not yet ripe. Instruments were too expensive, their transport was cumbersome, and their comparability was not assured. It was over half a century later that the first regular meteorological network was established. Its organizer was J. J. Hemmer, court chaplain to Karl Theodor, Prince Elector of the Palatinate. Hemmer persuaded the Elector to finance the purchase of a uniform set of instruments which he distributed with a set of detailed instructions and a uniform reporting scheme to clerical colleagues and correspondents at thirty localities. Of these, fourteen were in Germany, two in North America, and the remainder in other states in Europe. Many of them were monasteries and a few of these stations are still operating at the same location, notably Hohenpeissenberg in Bavaria and the Grand St. Bernhard in Switzerland.[14]

Hemmer also obtained the cooperation of several already existing meteorological observatories, including those in Cambridge, Massachusetts and Prague. The observations were regularly sent to the Societas Meteorologia Palatina and were published in yearly volumes as the *Ephemerides of the Society*. For a decade three daily observations were published until Hemmer's untimely death and the war turmoil put an end to this enterprise,

13 Karl August Wittfogel and Teng Chiao-Sheng, *History of Chinese Society: Liao, 907–1125* (Philadelphia, 1949), 752; Central Meteorological Research Bureau of China, "A Study of Occurrences of Flood and Drought of the Last 500 Years in Northern and Northeastern China (translated title)"; *Collected Papers on Climatic Change and Long Range Weather Forecasts* (Peking, 1977), 164–170; W. Wada, "Korean Rain-Gauges of the Fifteenth Century," *Quarterly Journal of the Royal Meteorological Society*, XXXVII (1911), 83–85; Arakawa, "On the Secular Variation of Annual Totals of Rainfall at Seoul from 1770 to 1944," *Archiv für Meteorologie Geophysik und Bioklimatologie*, VII (1956), 406–412. Christian Pfister, "Die älteste Niederschlagsreihe Mitteleuropas: Zürich 1708–1754," *Meteorologische Rundschau*, XXXI (1978), 56–62; J. M. Craddock, "Annual Rainfall in England since 1725," *Quarterly Journal of the Royal Meteorological Society*, CII (1976), 823–840.

14 Jacob Jurin, "Invitatio ad Observationes meteorologicas communi consilio instituendas," *Royal Society of London Philosophical Transactions* (1723), 422–427; Friedrich Klemm, "Johann Jacob Hemmer," *Neue Deutsche Biographie*, VIII (1968), 510–511.

although the observations continued at many places under Hemmer's scheme. Thus a number of localities can now boast of two centuries of instrumental meteorological observations.

The publication of the first decade of observations had far-reaching results for the science of meteorology. With them Humboldt constructed, together with some other data then available, the first isothermal map of the northern hemisphere in 1817. And a few years later the same data enabled Brandes to look synoptically at the observations and show that weather systems were indeed migratory and could be pursued across the map. It was the beginning of weather forecasting.[15]

In the early decades of the nineteenth century networks of weather stations were established in all major nations. The founding of the International Meteorological Organization in 1873 further stimulated the expansion of observation and established uniform rules. But there remained many regions of the world without regular weather observations. Only the expansion of aviation after World War II led to the gaps in the networks being filled. Finally, today, the assessment of weather and climate is on a firm basis, due in no small measure to surveillance by satellites.

Clearly, the information prior to the inauguration of governmental weather services, i.e., since the second half of the nineteenth century, is beset with problems. It has been cogently pointed out by others that the use of chronicle material for climatological purposes is fraught with uncertainties. This applies both to interpretation of phraseology and to dating. In all instances corroborating information must be sought. Yet with critical analysis the residual value of documentary material remains high for the pre-instrumental period if the descriptions refer directly to weather. Considerably less reliable are the inferences that can be drawn from references to crop yields. Modern models of crop ecology have shown how complex the response of crop plants is to weather. Sometimes even singular events of limited extent, such as a hailstorm, can adversely affect a crop. This applies particularly to fruits, including grapes. Crop failures because of drought can generally be accepted as evidence because

15 *Ephemerides Societatis Meteorologicae Palatinae* (Mannheim, 1784–1795), 12 v.; Alexander von Humboldt, "Des lignes isothermes et de la distribution de la chaleur sur le globe," *Mémoires de physique et chimie Societé d'Arceuil,* III (1817), 462–602; Heinrich Wilhelm Brandes, *Beiträge zur Witterungskunde* (Leipzig, 1820), 211.

such conditions are usually widespread. But in this, as in other cases, independent direct weather information, such as low water levels in rivers and streams or the drying out of springs and wells, is usually available.[16]

Inferences on weather drawn from crop prices are not acceptable because prices also reflect a number of other variables. Some valid deductions can often be drawn from references to the quantity, quality, and harvest date of grapes and wine, but even that is fraught with difficulties. Similarly, interpretation of advances and retreats of glaciers is not unambiguous. The common deduction that advancing mountain glaciers are synonymous with colder climate is not necessarily supportable. In some regions warm winters with increased snowfall can be responsible. At times glaciers at various elevations in a mountain range can exhibit simultaneously opposite behavior.[17]

One of the shortcomings of the sources of weather information cited above is that they are dominated by material from Central and Western Europe. These data cannot be extrapolated far beyond their origin or even be considered to represent hemispheric or global conditions. Yet the long observational series for central England, DeBilt, Berlin, Kiel, and St. Petersburg can be used to derive a picture of cold winters in Central Europe to about 1679. Figure 1 shows the departures of temperature from the average in extremely cold winter months in that area between 1679 and 1978. Only months with two standard deviations or colder than the average are shown. The outgoing seventeenth and the first half of the eighteenth century had few exceptionally cold winters. The high frequency of such very cold winter months in the interval from 1767 to 1855 is notable. Just as interesting is the near-absence of very cold winter months between 1856 and 1928.[18]

16 W. T. Bell and A. E. J. Ogilvie, "Weather Compilations as a Source of Data for the Reconstruction of European Climate during the Medieval Period," *Climatic Change,* I (1978), 331–348; Martin J. Ingram and David J. Underhill, "The Use of Documentary Sources for the Study of Past Climates," paper delivered at the International Conference on Climate History (Univ. of East Anglia, 1979), 59–90.

17 Emmanuel Le Roy Ladurie (trans. Barbara Bray), *Times of Feast, Times of Famine: History of Climate Since the Year 1000* (Garden City, 1971), 426.

18 Rudolf Fischer, "Die Kältesten Wintermonate in Berlin 1719–1941," *Zeitschrift für Angewandte Meteorologie,* CX (1943), 375–376; Hans von Rudloff, *Die Schwankungen und Pendelungen des Klimas in Europa seit Beginn der regelmässigen Instrumenten-Beobachtungen (1670)* (Braunschweig, 1967), 370.

Fig. 1 Extremely Cold Winter Months 1679–1978, C. Europe
~ Departure, °C
(↓ 2 Extremely Cold Months in Same Winter)
(– – – Estimated)

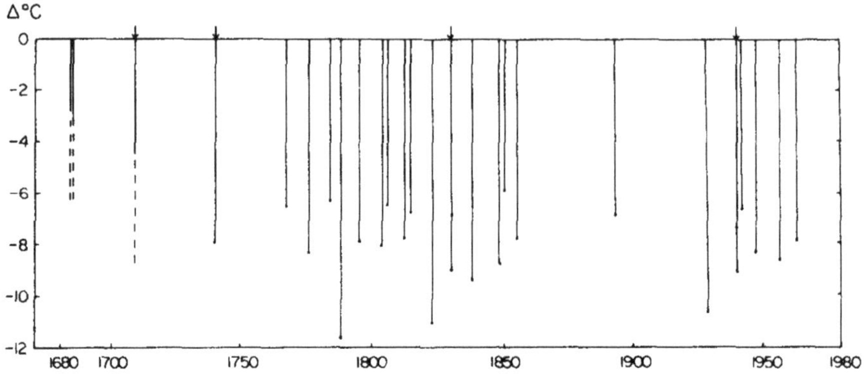

It must be emphasized here that the old thermometer readings need to be carefully interpreted. This is not an easy task. Aside from imperfections of the capillaries in the thermometers, there are ambiguities of scales and calibration, and hours of observation, and uncertainties about exposure of the instruments. With over forty thermometer scales in use during the eighteenth century there were problems of comparability. Only in the first decades of the nineteenth century was some standardization introduced.[19]

The question arises: Can one infer something about the global climatic fluctuations in the last few centuries from the data now available? With the lack of information from the southern hemisphere the answer is *no*. We are somewhat better off with respect to the northern hemisphere. Using fairly complete data sources for the past century, Groveman and Landsberg have devised a multiple regression scheme which permits reconstruction of northern hemisphere temperature departures from the established 1881–1975 mean value. The procedure makes use of observational

19 George Martine, *Essays and Observations on the Construction and Graduation of Thermometers and on the Heating and Cooling of Bodies* (Edinburgh, 1787; 4th ed.), 177; J. J. Luz, *Vollständige und auf Erfahrung gegründete Anweisung Thermometer zu verfertigen* (Nürnberg, 1781); Kristine Meyer, *Die Entwicklung des Temperaturbegriffs in Laufe der Zeiten* (Braunschweig, 1913), 160; Landsberg, "A Note on the History of Thermometer Scales," *Weather*, XIX (1964); W. E. Knowles Middleton, *A History of the Thermometer and its Use in Meteorology* (Baltimore, 1966), 249.

Fig. 2 Reconstruction of Northern Hemisphere Temperature Departures from 1881–1975 Mean for the Period from 1579 to 1880.[a]

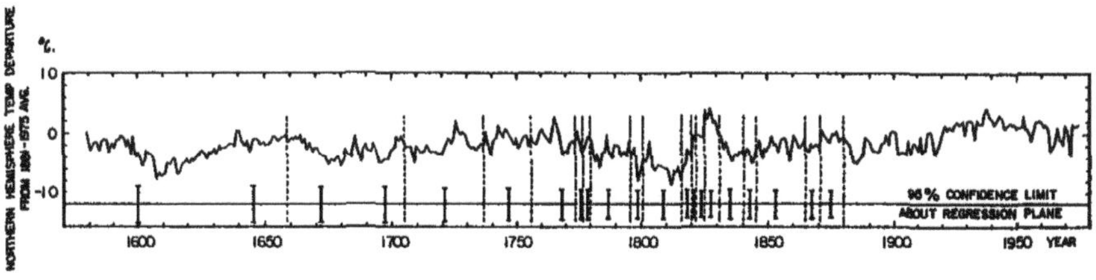

a Dashed vertical lines indicate various intervals for which separate regression equations were used for the reconstruction. Confidence limits are indicated below the curve.

temperature series, especially at high latitudes, and some proxy data such as Lake Suwa's freezing dates and several series of tree ring observations to simulate conditions. The results for the interval from 1579 to 1880 are shown in Figure 2.[20]

There are clearly no hemispheric temperature trends in that interval but rather irregular ups and downs. There is little justification to label this whole interval the little ice age. There are, however, some persistent cold intervals. Very prominent is the cold decade from 1605 to 1615 and other cold spans from 1674 to 1682 and from 1695 to 1698. The first half of the eighteenth century was relatively warm, even though the years 1709 and 1740 seem to have been cold all over the hemisphere, not just in Europe. From about 1770 onward there was a fairly steady drop in temperatures until about 1820. This drop corresponds to an interval with increasing snowfall in Scotland. The years from 1805 to 1820 were the coldest era in Europe, eastern North America, and Japan. This period includes the notoriously cold year of 1816, the year without a summer in eastern North America and Western Europe. This cold spell has often been attributed to the massive Tambora eruption in 1815, which may well be an erroneous interpretation because notable hemispheric cooling had already started a decade earlier. Actually 1812 was probably the coldest year in the last four centuries. However, the possibility of a volcanic dust veil as a cause for the five cold decades cannot be entirely ruled out. From 1825 to 1830 there was a notable warming to be followed again by cooling. The last three decades of the

20 Brian S. Groveman and Landsberg, *Reconstruction of Northern Hemisphere Temperature: 1579–1880* (College Park, 1979).

nineteenth century were quite cold in parts of the northern hemisphere with a notable shift of the center of the cold toward the Far East. In Japan from 1866 to the end of the century many cold seasons occurred. Chinese scholars have noted that the interval from 1876 to 1895 lacked warm seasons even in the south of China, with temperatures below average.[21]

There have been studies which advocated that the dearth of sunspots from 1645 to 1715, the so-called Maunder minimum, resulted in a stretch of low temperatures. Our reconstruction does not bear this out. The coldest period started six decades later, when solar activity seems to have been normal.[22]

More insight on this and other questions of climatic history over the past few centuries may be obtained by systematic collection and analysis of information in diaries and other historical documents. The re-awakened interest in climate and in the impact of climatic fluctuations on human activities makes it desirable to uncover these sources of information and evaluate the material for the disclosure of weather events and climatic trends.

21 See for example, Mowbray G. Pearson, "The Winter of 1739–40 in Scotland," *Weather*, XXVII (1973), 20–24; *idem*, "Snowstorms in Scotland, 1782–1786," *ibid.*, XXVIII (1973), 195–201. Landsberg and Jesselle M. Albert, "The summer of 1816 and volcanism," *Weatherwise*, XXVII (1974), 63–66; Change Chia-cheng, Wang Shao-wu, and Cheng Szu-chung, "Climatic Change and the Exploitation of Climatic Resources in China," paper delivered at the World Climatic Conference (Geneva, 1979), 15.

22 John A. Eddy, "Climate and the Changing Sun," *Climatic Change*, I (1977), 173–190; Landsberg, *Some Previously Unrecorded Observations of Sunspots during the 'Maunder Minimum' and Auroras not Listed in Standard Catalogs* (College Park, 1979).

Andrew B. Appleby

Epidemics and Famine in the Little Ice Age

Mortality in Western Europe during the sixteenth and seventeenth centuries, as Flinn has shown, was characterized by great instability: periods of "crisis" mortality alternating with periods of relatively low mortality. The frequent crises were caused by famine, epidemic disease, and war, sometimes working in combination, sometimes not. After about 1650 or 1700, these great crises became progressively less frequent and mortality moved slowly and unevenly toward its present state of stability, with little fluctuation in crude death rates from year to year.[1]

Famine, the most important cause of these early modern mortality crises, disappeared from much of Western Europe by the early eighteenth century. Southern England suffered its last major food crisis—which reached famine proportions in certain regions—in 1597. Northern England experienced another wrenching famine in 1623 and a slight one in 1649 but weathered the harvest failures of the 1690s without widespread starvation. A few parishes in the English Midlands were struck by famine from 1727 to 1728, but thereafter England has been free of famine. Scotland had terrible famines in 1623 and again in the 1690s but then managed to avert widespread starvation after harvest failures. France was especially subject to famine in the seventeenth and early eighteenth centuries, with terrible crises falling in 1630–1631, 1649–1652, 1661–1662, 1693–1694, and 1709–1710. After 1710, most of France was famine-free until 1795, when the dislocations of war and the confusion of the government again brought starvation to the poor. Germany, Scandinavia, and Switzerland suffered famine from 1770 to 1772 and many parts of Europe experienced yet another food crisis in 1816. Ireland, of course, was devastated by famine as late as the 1840s.[2]

The author is indebted to John Post for comments on his article and has also benefited greatly from the criticisms of Mary Schove Dobson.

1 Michael W. Flinn, "The Stabilisation of Mortality in Preindustrial Western Europe," *Journal of European Economic History*, III (1974), 285–318.

2 Appleby, *Famine in Tudor and Stuart England* (Stanford, 1978), 121–156; A. Gooder, "The Population Crisis of 1727–30 in Warwickshire," *Midland History*, I (1972), 1–22; Michael Flinn et al., *Scottish Population History from the 17th century to the 1930s* (Cambridge,

This brief chronology suggests that famine disappeared from Western Europe gradually and unevenly and that extremely poor harvests—as in 1816—could threaten the poor with starvation as late as the nineteenth century. Even more recently, wars have brought starvation, as in Holland in 1944–1945 and in Warsaw in 1941–1943. These last examples, however, can be considered "man-made" famines and, putting them aside, Flinn is undoubtedly correct in concluding that "at some point . . . between the late seventeenth and late eighteenth centuries, according to chronologies that varied from country to country and region to region, harvest failures generally ceased to be seriously, or at least consistently, lethal."[3]

As famine retreated from Western Europe, so too did the great epidemic diseases. Plague disappeared from Western Europe in the middle years of the seventeenth century, except for occasional outbreaks such as the Marseilles epidemic from 1720 to 1722. Typhus became less common after about 1648, although it could re-emerge in epidemic form in wartime or during acute food shortage. Smallpox, however, became more common in the eighteenth century than before, at least in England.[4]

The rise and decline of other diseases are difficult to trace because of problems in identifying them but there is evidence that childhood maladies, particularly infantile diarrhea, were also on the decline in the eighteenth century. Infantile and child mortality in London, for example, increased during the late seventeenth and early eighteenth centuries and then slowly subsided. Infantile diar-

---

1977), 164–186; Jean Meuvret, "Les crises de subsistances et la démographie de la France d'Ancien Régime," *Population*, I (1946), 643–650; Pierre Goubert, *Beavais et le Beauvaisis de 1600 à 1730* (Paris, 1960), 45–58, 75–80, 300–303; François Lebrun, *Les hommes et la mort en Anjou aux 17e et 18e siècles* (Paris, 1971), 131–138, 329–346; Jacques Dupâquier, "Sur la population française au XVIIe et au XVIIIe siècle," *Revue historique*, CCXXXIX (1968), 66; Richard Cobb, *Terreur et subsistances* (Paris, 1965), 307–342; Wilhelm Abel, *Massernarmut und Hungerkrisen im vorindustriellen Deutschland* (Göttingen, 1972), 46–54; Christian Pfister, "Climate and Economy in Eighteenth-Century Switzerland," *Journal of Interdisciplinary History*, IX (1978), 223–243; John D. Post, *The Last Great Subsistence Crisis in the Western World* (Baltimore, 1977), passim.

3 Flinn, "Stabilisation of Mortality," 302.

4 Jean-Nöel Biraben, *Les hommes et la peste en France et dans les pays européens et méditerranéens: La peste dans l'histoire* (Paris, 1975), I, 336–337, 388, 394, 399–400, 407; Post, *Last Great Subsistence Crisis*, passim; August Hirsch (trans. Charles Creighton), *Handbook of Geographical and Historical Pathology* (London, 1883), I, 578–580; Peter Razzell, *The Conquest of Smallpox* (Firle, Sussex, 1977), 113–139.

rhea followed a different course in France, where it was unusually prevalent toward the end of the eighteenth century.[5]

The frequency of great epidemics taken as a whole, however, declined around the end of the seventeenth century, coincident with the lessening impact of harvest failures. The gradual disappearance of dramatic epidemics did not, by itself, bring any immediate improvement in life expectancy. Rather there was a change in the type of disease from the occasional great epidemic to a high, more constant level of mortality. In other words, the crisis mortality was flattened but background mortality rose and it was only toward the end of the eighteenth century that life expectancy improved in France and after the beginning of the nineteenth century that life expectancy increased in England.[6]

Local variations in timing and geography aside, the period from about 1550 to 1700 was the age of great mortality crises, at least when compared to the following centuries. We know less about how this period compares with earlier centuries but certain—not all—epidemic diseases were more widespread after 1550 than before and famines were also more frequent and possibly more terrible after 1550 than at anytime since the early fourteenth century.

This period of crisis coincides neatly with the dates that Lamb has chosen for the "little ice age." Between 1550 and 1700, Lamb notes, temperatures were unusually cold and the climate was characterized by great instability. The purpose of this article is to consider whether the cold and variable climate of the little ice age contributed to the crisis mortality and whether the warming period around 1700 can be related to the decline of crisis mortality that Flinn has traced.[7]

5 John Marshall, *Mortality of the Metropolis* (London, 1832); J. P. Goubert, "Le phénomène épidémique en Bretagne à la fin du XVIIIe siècle," *Annales,* XXIV (1969), 1562–1588; François Lebrun, "Les épidémies en Haute-Bretagne à la fin de l'Ancien Régime (1770–1789)," *Annales de démographie historique* (1977), 181–206.

6 Yves Blayo, "La mortalité en France de 1740 à 1829," *Population,* numéro special (1975), 123–137; Roger Schofield and E. A. Wrigley (eds.), *Population Trends in Early Modern England,* forthcoming.

7 H. H. Lamb, *Climate: Present, Past and Future: Climatic History and the Future* (London, 1977), II, 463. It became obvious from the discussion at the Climate and History Conference (Cambridge, Mass., 1979), that the dating of the "little ice age" is uncertain and that the concept of a "little ice age" may not be useful.

In many ways, the connection between infectious diseases and climate is very close. Broadly speaking, diseases are transmitted from one person to another either through the air, by water, or by insects, such as lice, fleas, flies, or mosquitoes. At least theoretically, under extreme climatic conditions many diseases cannot spread because the pathogens themselves—and in the case of insect-borne diseases, the insect carriers—are vulnerable to excessive heat or cold, wetness or dryness. Many airborne infections, for example, are hindered by hot, sunny weather, which dries and kills the airborne microorganisms before they can enter a new host.

It has been suggested, for instance, that when the ancestors of the American Indians crossed the Bering Straits, the cold prevented many diseases, including smallpox, that were present in Asia from being introduced into the Americas—with disastrous future consequences for the descendants of these migrants. Disease-carrying insects are also vulnerable; *Xenopsylla cheopis,* one of the most important flea carriers of the plague, flourishes only when the temperature is between 68 and 78 degrees Fahrenheit and there is some humidity. The fleas' eggs do not hatch below 55 degrees and temperatures below 45 degrees kill the eggs. Ranges of temperature and humidity determine the plague season, which is summer and autumn in temperate climates and winter in torrid zones. Epidemic infantile diarrhea, which was carried by flies, was characteristically a late summer disease, when in temperate climates flies are most numerous. The geographical distribution of the malaria mosquito is also controlled by temperature, although the mosquito can range as far north as Archangel in European Russia. Malaria was once an important disease in northern Europe, including England, and may have accounted for substantial numbers of victims in the somewhat mysterious epidemics of recurring fever that occasionally swept across northern Europe. Many other diseases follow seasonal patterns that strongly suggest a close tie between disease transmission and certain favorable climatic conditions.[8]

8 Marshall T. Newman, "Aboriginal New World Epidemiology and Medical Care, and the Impact of Old World Disease Imports," *American Journal of Physical Anthropology,* XLV (1976), 667–672; L. Fabian Hirst, *The Conquest of Plague* (Oxford, 1953), 260–276; G. Melvyn Howe, *Man, Environment and Disease in Britain* (New York, 1972), 132; John Brownlee, "The Health of London in the Eighteenth Century," *Proceedings of the Royal*

Climate can also influence disease transmission in a subtler way by favoring mutant forms of pathogens while acting unfavorably on the older form of the same pathogen. It is likely, for example, that venereal syphilis arose as a mutant form of another spyrochitic disease (either yaws or endemic childhood syphilis) during the warming period of the later Middle Ages. According to this theory, yaws—identified within a broad category of diseases then called leprosy—was common in the early Middle Ages. It was transmitted through skin contact when inadequately clothed people huddled together for warmth. With the coming of a colder period and the increased use of clothing in the second half of the fourteenth century, prolonged skin contact became less common and the incidence of the disease declined. A mutant form that depended on venereal contact appeared, however, at the end of the fifteenth century and thrived, for a short time, in its new syphilitic form. The disappearance of leprosy and the appearance of syphilis thus may be explained in part by climatic change.[9]

Climate thus is an important factor in disease transmission. But what is the connection between the little ice age and the unusual frequency of great epidemics? We might suppose that disease increased in importance at the beginning of the cooling, unstable period—that is, around 1550—and declined when the weather turned warmer around 1700. To test this possibility, we consider some of the major epidemic diseases of the period.

Plague, the most cataclysmic of all epidemic diseases, disappeared from Western Europe in the mid-seventeenth century, in the middle of the little ice age. Following a series of massive epidemics that ravaged northern Italy and southern France from 1629 to 1631, southern Italy and Genoa from 1656 to 1657, and northern France, England, and Holland in the 1660s, Western Europe became plague-free, except for the isolated epidemic of 1720–1722 in southern France. Lamb has noted that the great

---

Society of Medicine, XVIII (1925), 73–85; L. W. Hackett, "Conspectus of Malaria Incidence in Northern Europe, the Mediterranean Region and the Near East," in Mark F. Boyd (ed.), *Malariology* (Philadelphia, 1949), II, 793; Schofield, "An Anatomy of an Epidemic: Colyton, November 1645 to November 1646," in *The Plague Reconsidered* (Matlock, Derbyshire, 1977), 121.

9 C. J. Hackett, "On the Origin of the Human Treponematoses (Pinta, Yaws, Endemic Syphilis, and Venereal Syphilis)," *Bulletin of the World Health Organisation*, XXIX (1963), 7–41; William McNeill, *Plagues and Peoples* (New York, 1976), 175–180.

plague pandemics appeared in times of unusual climatic instability and he implied that this instability affected the incidence of plague.[10]

But it is difficult to see any connection between the climatic instability and plague's occurrence. Plague appeared in the mid-fourteenth century, during a cool and unstable period, but it then recurred regularly through relatively warm and stable periods, as it did between 1500 and 1550, before disappearing during yet another cool, unstable period in the mid-seventeenth century. The reasons for the plague's eventual disappearance from Western Europe are a matter of debate. The most widely accepted explanation hinges not on climatic change but on improvements in methods of quarantine.[11]

Indeed, the cooling of the little ice age should have slowed plague activity and reduced the number of epidemics, if the cooling were marked by lower than usual summer temperatures. In the temperate zones of Europe, cool summers would have retarded the breeding activity of fleas and lessened chances of an epidemic. Biraben, however, has shown that plagues were frequently reported from 1550 to about 1670, that is, during the little ice age.[12]

I am not implying that climate had no effect on plague. Unfavorable climatic conditions—such as a very dry, cold winter or a dry, hot summer—brought to a halt epidemics that had already begun. Favorable weather alone, however, was not enough to cause epidemics, even in those areas where the disease was present endemically. Plague was endemic in London for most of the sixteenth and seventeenth centuries and epidemics often came in years when the climatic conditions were favorable. According to Schove, the epidemics of 1603 and 1665 coincided with favorable weather. The epidemic of 1625 did not, however, and apparently in 1630 the weather was suitable but no epidemic

10 Biraben, *Les hommes et la peste*, I, 388, 393–394, 399–400, 407–421; Lamb, *Climate*, II, 262; David J. Schove, "Chronology and Historical Geography of Famine, Plague and other Pandemics," *Proceedings of the XXIII Congress of the History of Medicine* (London, 1972), 1265–1272.

11 Biraben, *Les hommes et la peste: Les hommes face à la peste* (Paris, 1976), II, 169–175; Flinn, "The Disappearance of the Plague," *Journal of European Economic History*, VIII (1979), 131–148. For a different explanation, see Appleby, "The Disappearance of the Plague: A Continuing Puzzle," *Economic History Review*, forthcoming.

12 Biraben, *Les hommes et la peste*, I, 132.

followed in London. In other words, climate was one of the factors influencing the timing of epidemics but only one. The others, that probably were more important, remain obscure.[13]

Plague is not a good disease to relate to short-term climatic change. We know too little about either temperature or humidity during the time that plague was present in Western Europe to permit us to draw any firm conclusions about the recurring nature of the disease and the climate.

The climatic records are better during the period of major smallpox activity. Smallpox was a relatively rare disease—in England, if not in southern Europe—until the 1630s. It then became more prevalent and after 1666 replaced plague as the most feared of all diseases. The fear of smallpox seems to have been due in large part to its loathsome nature; in London, where mortality from various diseases can be compared, it was by no means as great a killer as either "consumption," a catch-all term that covered tuberculosis and all other wasting diseases, or "convulsions."[14]

The little ice age had little to do with the increased importance of smallpox. The disease waxed during the cold of the seventeenth century and then remained active throughout the warmer eighteenth century, only to die down with the adoption of inoculation and vaccination. Climate also was unimportant in the timing of epidemics over the short term. Smallpox was endemic among the children of London for most of the seventeenth and eighteenth centuries. Every few years, however, there was a surge in smallpox deaths; thirty-seven epidemics can be identified in the years from 1660 to 1799, or about one every 3.78 years. (Epidemics are here defined as those years when smallpox accounted for over 10 percent of all deaths in London, as reported in the Bills of Mortality.)[15]

These recurring epidemics do not fit any particular climatic pattern, either in temperature, for which we have figures from 1660, or in rainfall, for which we have evidence from 1697.

13 *Ibid.*, 118–154; Schove, "Chronology," 1271; Biraben, *Les hommes et la peste*, I, 116; T. Birch (ed.), *A Collection of the Yearly Bills of Mortality from 1657 to 1758 inclusive* (London, 1759).

14 Thomas R. Forbes, *Chronicle from Aldgate* (New Haven, 1971), 100; Appleby, "Nutrition and Disease: The Case of London, 1550–1750," *Journal of Interdisciplinary History*, VI (1975), 20–22.

15 Marshall, *Mortality of the Metropolis*, passim.

Epidemics came in very dry years, as in 1731, or in wet years, as in 1768. They broke out in the warm years of 1736, 1749, and 1779, and in unusually cold years, such as 1740, which was the coldest of the entire period.[16]

However, severe winters, such as that in 1740, did not invariably trigger epidemics of smallpox. In the even colder winter of 1684—the worst ever recorded in southern England—no epidemic took place in London. During the entire decade of the 1690s, which had no fewer than six winters with mean temperatures below 3 degrees centigrade, London suffered no smallpox epidemics whatever, although the disease smouldered endemically in the city throughout that decade.

Smallpox thus does not seem to have been controlled by the weather, according to my cursory analysis. I may have chosen the wrong climatic criteria; a close statistical investigation of more variables than I have chosen might turn up a significant relationship between epidemic outbreaks and certain types of weather. It is unlikely, however, that the long-term rise and decline of smallpox can be traced to secular changes in the climate.

The greatest killer of late seventeenth- and early eighteenth-century London was "convulsions." This heading in the bills of mortality refers not to a disease but rather to a terminal symptom. Convulsions in children—and almost all deaths from convulsions were of infants and small children—can arise from a wide variety of different diseases. The major component of this category of mortality in the London bills, however, was infantile diarrhea. Linked to convulsions was "griping in the guts," another leading cause of child mortality in seventeenth-century London, which also describes a symptom, rather than a disease. The number of deaths attributed to griping declined steadily in the 1690s and became insignificant by the 1740s. As griping became less important, the mortality from convulsions grew, but the decline of one disease and the coincidental rise of another was more apparent than real. Creighton has argued convincingly that both griping

16 Gordon Manley, "Central England Temperatures: Monthly Means 1659 to 1973," *Quarterly Journal of the Royal Meteorological Society*, C (1974), 389–405; B. G. Wales-Smith, "Monthly and Annual Totals of Rainfall representative of Kew, Surrey, from 1697 to 1970," *Meteorological Magazine*, C (1971), 345–362; F. J. Nicholas and J. Glasspoole, "General Monthly Rainfall over England and Wales 1727 to 1931," *British Rainfall* (1931), 299–309; all reproduced in Lamb, *Climate*, II, 572–574, 620–622.

in the guts and convulsions were names given to infantile diarrhea and those deaths that appeared at one time under the heading of griping were at a later period entered under the rubric of convulsions. In other words, what appears as a decline in griping was merely a transfer of the same underlying disease to another heading.[17]

In this analysis of convulsions, I have therefore combined the mortality from griping and convulsions since they were both made up primarily of mortality from infantile diarrhea. It is possible, with the two categories combined, to trace the course of infantile diarrhea in London. The disease first became seriously epidemic in 1669 and 1670, then retreated for a few years, but increased again in the 1680s. It remained at a very high level until the 1730s, when it began to decline, although it was still an important killer of the very young until World War I. The period of highest mortality—from 1669 through 1732—was at the end of the little ice age and continued into the warmer eighteenth century, suggesting that the overall coolness of the ice age itself had little influence on the long-term prevalence of the disease. One might expect, however, a strong short-term correlation between the timing of epidemics and climatic conditions. Infantile diarrhea was associated in the minds of contemporaries with unusually warm, dry summers and mortality from this disease was concentrated in the late summer and early autumn.[18]

Although mortality from infantile diarrhea was relatively stable from year to year, certain years were epidemic in that mortality from convulsions and griping was higher than usual or represented a greater proportion than usual of total London deaths. These epidemic years were 1669, 1670, 1676, 1699, 1702, 1704, 1705, 1718, 1723, 1726, and 1733. In 1669 and 1676 summer temperatures were appreciably above normal, with normal defined as the average summer temperatures from 1659 to 1749. The summer of 1670 was in the normal range. Rainfall figures are not available for these years. The climate of 1699 was apparently ideal for summer diarrhea, if it indeed depends on climatic

17 Thomas Willis, *The London Practice of Physick* (London, 1685), 250; Charles Creighton, *A History of Epidemics in Britain* (Cambridge, 1894), II, 747–756.

18 Brownlee, "The Health of London," 77–78; Walter Harris, *An Exact Enquiry Into, and Cure of the Acute Diseases of Infants* (London, 1693), 39; Creighton, *Epidemics in Britain*, II, 749–751.

conditions for its prevalence. The spring and summer of 1699 were both drier than the average for the years 1659 through 1749 and the summer was slightly warmer than usual. The next epidemic summer, 1702, was moderately dry but the temperature was a little below normal. In 1704, the summer was dry and warm and in 1705 the weather was again dry, although the temperature was just below normal. The next epidemic, that of 1718, saw a dry but slightly cool summer. Again in 1723 the summer was dry and the temperature was slightly above normal.[19]

To this point, the only correlation that can be found is between a dry summer and epidemics; three epidemics had fallen in years of below average temperature. In the next two epidemics—of 1726 and 1733—the summers were much wetter than normal, although both were warmer than normal and evaporation of accumulated surface water may have been high. If we look at those years when mortality was below normal, the evidence is equally inconclusive. The years 1700, 1709, 1711, 1725, and 1744 all saw unusually low mortality. The summers of 1700, 1725, and 1744 were cool, but those of 1709 and 1711 were warm. (The winter of 1709 was very severe and may have killed many of the eggs or grubs of flies that are thought to be the carriers of the disease.) The summers in 1709, 1711, and 1725 had been wet but 1700 and 1749 were dry. In short, no relationship between climatic conditions and epidemic activity in these childhood ailments can be found, either in a positive or negative sense, despite the observations of contemporaries that epidemic summers were hot and dry. Here again, however, caution is needed since the evidence may be too rudimentary to establish a relationship. The temperature and rainfall measures that we have would obscure particular favorable short-term weather conditions. It may have been, for example, that flies needed only a few days of warm, dry weather at a certain time of the summer to have bred profusely and that temperature and rainfall after this short period did not affect them. The published data, however, are too crude to allow a closer analysis.[20]

19 Lamb, *Climate*, II, 572–573.

20 Mary Schove Dobson has pointed out that London is not the best place to test the relationship between climate and the incidence of disease, since London had a unique environment which would probably have obscured any statistical relationship between climate and disease. She is currently examining differences in mortality in rural localities with different climates.

The relationship between malaria and climate is difficult to establish. The *Anopheles* mosquito requires July temperatures over 16 degrees centigrade to be infective and flourishes only when the temperature reaches 19 to 20 degrees. In northwestern Europe, including England and much of Scandinavia, epidemics broke out in warm summers in the sixteenth and seventeenth centuries, according to Lamb. The chilly summers of the 1690s eliminated the malady but it returned again in the warmer years of the eighteenth and especially the nineteenth centuries.[21]

That malaria only reached epidemic levels in warm years is likely, but pinpointing the dates of malaria epidemics is impossible because the disease was not differentiated from other recurring fevers. Malaria may have contributed to the enormous death tolls in the great, mysterious fevers of 1557–1558 and 1727–1730, although the first has been rather questionably identified as an epidemic of influenza and the latter was almost certainly a combination of several fevers following one another within a few years. Possibly the cold of the little ice age reduced the numbers of mosquitoes and the incidence of the disease between 1550 and 1700. The eventual decline of malaria in England and most of the rest of northwestern Europe seems, however, to have been a result of agricultural changes, in particular the increase in cattle raising (mosquitoes feed on cattle), and does not appear to have been related to climatic changes.[22]

Given the rudimentary evidence available to me, I have argued that climate has not been a major factor in either the long-term rise and decline of certain important epidemic diseases or in the timing of individual epidemics. Epidemic disease, however, cannot always be separated from harvest failures, which were the result of bad weather. When the harvest failed—or food prices soared for whatever reason—many poor people took to the roads in search of food or charity. They crowded into towns or cities where charitable resources were greater than those in the country;

21 Lamb, *Climate*, II, 188, 261; Howe, *Man, Environment and Disease*, 42–43.

22 F. J. Fisher, "Influenza and Inflation in Tudor England," *Economic History Review*, XVIII (1965), 120–129; Thomas Short, *New Observations on City, Town and Country Bills of Mortality* (London, 1750), 91; Hackett, "Malaria Incidence," 791. Dobson informs me that the geographical distribution of the fevers of 1727–1730 did not coincide with those regions where malaria was prevalent. She has also pointed out that malaria became more benign—or humans more resistant—within the last century and that unsuspected cases have turned up in England in which the person with the disease had no other symptoms than an occasional chill.

or, if a city closed its gates against the flood of vagrants, they pushed into the overcrowded slums outside the city walls. Diseases that were present in non-epidemic form blossomed into epidemics and were diffused from one locality to another.

For example, the English counties of Cumberland and Westmorland suffered a frightful famine in 1597. To obtain food, many local people journeyed across the Pennine mountains to Newcastle in Northumberland where some food was available—but where the plague was raging. The next year, 1598, plague broke out in Cumberland and then spread to Westmorland, probably brought into the area by the people who braved the disease to buy food the previous year. In this way, harvest failure and the weather conditions that caused it brought, at one remove, an epidemic to these two counties.[23]

Many scholars have argued that the malnutrition which followed harvest failures favored the outbreak of epidemics—or increased their deadliness—because malnourished bodies offered less resistance to disease. This theory is plausible, but it has not yet been proven. A number of non-conforming instances have even been found. In London, for example, where most diseases were endemic during the early modern period, epidemic outbreaks have been surprisingly independent of the price of food-stuffs and the level of nutrition of the poor. Even outside London, higher disease mortality cannot be found following all harvest failures. In the 1690s, for instance, repeated poor harvests pushed English wheat prices to very high levels, although there was evidently no starvation. At the same time, unemployment rose, both from the shift in demand away from manufactured goods toward food-stuffs and from depressions in certain segments of the economy caused by war. Put crudely, the "dear" 1690s should have been the "sick" 1690s, if disease flourishes in underfed bodies. But all across England mortality was at a low level. During the 1680s, on the other hand, grain prices were low and the poor should have been well-fed. Yet disease was rampant.[24]

23 Appleby, *Famine,* 109, 113.

24 Macfarlane Burnet, *Natural History of Infectious Diseases* (Cambridge, 1953), 142–143; Nevin S. Scrimshaw, C. E. Taylor, and J. E. Gordon, *Interactions of Nutrition and Infection* (Geneva, 1968), passim; Michael C. Latham, "Nutrition and Infection in National Development," *Science,* CLXXXVIII (1968), 561–565; Appleby, "Nutrition and Disease," 1–22; J. D. Chambers, *Population, Economy and Society in Pre-Industrial England* (Oxford, 1972), 30–31, 77–106; Ransom Pickard, *The Population and Epidemics of Exeter in Pre-Census Times* (Exeter, 1947), 61, 67; Cambridge Group data.

Among all age groups, children are the most vulnerable to prolonged malnutrition and one might expect an upturn in infant and child mortality after harvest failure even if adult mortality remained low. The London Bills of Mortality, where child mortality can be separated from adult mortality, suggest that the very young may, indeed, have suffered from the food shortages of the 1690s. Mortality from griping in the guts and convulsions was slightly elevated in 1694, when the effects of the harvest failure of 1693 would have been felt, and it reached epidemic levels in 1699, after the harvest failures of 1697 and 1698. Possibly 1694 would have seen greater mortality had the summer not been abnormally cold. The summer of 1699 was, as I remarked earlier, both warm and dry and possibly favored an epidemic of infantile diarrhea, which in turn may have been especially deadly because many children were malnourished that year.

Not too much of a connection, however, can be made between childhood ailments and high prices. Of the ten epidemics of infantile diarrhea only two—those of 1699 and 1726—came in years of high grain prices. The 1690s were without any epidemics of smallpox, which was also predominantly a childhood malady. In short, disease could be spread by vagabondage—and probably by the grain trade—following harvest failures but food shortage did not invariably result, even among children, in increased disease mortality. This is not to deny that people died from starvation, but it was often unaccompanied and uncomplicated by epidemic disease.

Even when epidemic disease was not present, harvest failures sometimes brought enormous mortality through starvation. In Penrith, a small town in Cumberland, famine in 1597 pushed mortality to over four times normal, and another famine in 1623 raised the town's mortality to more than five times normal. (Normal is here defined as the average number of burials in a series of non-crisis years.) If the normal death rate was thirty per thousand per year, these famines killed, respectively, 9 and 12 percent of the town's population. (These figures may be inflated by the deaths of immigrants who came to the town seeking charity and died there.) Ferocious as these death rates were, they sink to insignificance beside the mortality from plague in 1598, which was an unexpected by-product of the famine of 1597. Plague in 1598 carried off about 35 percent of the population, with a mortality almost thirteen times normal. The great famines of France at the end of the seventeenth century took tolls similar to those found in northern England a century earlier. Goubert has estimated, for example, that 10 percent of the population of northern France died in 1693–1694; in the Auvergne, the figure may have reached 20 percent. Famine in its own right was a terrible killer.[25]

To what degree were these early modern famines caused by adverse weather? And to what degree were they caused by economic and social weaknesses, unrelated to the weather?

Historians have usually assumed that famine mortality was a direct result of harvest failures which were in turn the result of bad weather. With demand more or less constant—there being no substitute for grains before the introduction of the potato—the reduced supply meant higher prices. When grain prices reached certain heights, or remained high for a prolonged period, the poor were unable to buy the food necessary to sustain life and they died—from starvation, from eating unsuitable food substitutes such as tree bark or grass, or from diseases directly related to malnutrition, such as the "bloody flux." The initial step in this causal sequence was the weather and the entire process can be baldly summarized as "bad weather caused starvation, good weather brought cheap food."

In part, this was true. Besides disruptions from war or civil unrest the weather was certainly *the* major short-term factor in the quantity and quality of the harvest. Bad weather often seriously reduced the harvest yields over large parts of Europe at the same time, as in 1693, 1709, 1739–1740, 1770, and 1816. The unseasonably cold summer of 1816 was also felt in North America. The earlier European harvest failures of 1596 and 1630 were paralleled in India, which suffered famines in both those years. The wide geographical distribution and similarity in timing of these bad harvests suggest that only climate could have been the culprit, in view of the varied economic and social conditions in the afflicted regions. Thus the first step in the process is clear enough: unfavorable weather was the primary cause of harvest failure.[26]

25 Appleby, *Famine*, 224, 1n.; Goubert, *Louis XIV et vingt millions de Français* (Paris, 1966), 166–170; A.-G. Manry (ed.), *Histoire de l'Auvergne* (Toulouse, 1974), 304.

26 For an example of the dreadful effects of war, see Jean Jacquart, "La Fronde des Princes dans la region parisienne et ses conséquences matérielles," *Revue d'histoire moderne et contemporaine*, VII (1960), 257–290. Post, *Last Great Subsistence Crisis*, 7–14; Fernand Braudel (trans. Miriam Kocham), *Capitalism and Material Life 1400–1800* (New York, 1975), 41.

However, the relationship between the size of the grain harvest and the price of grains is not so unambiguous. Little has been known until recently about harvest yields in pre-industrial Europe and it has usually been assumed that grain prices faithfully reflected the quantity of the grains harvested. Hoskins, for example, assigned English harvest qualities on the sole criteria of price: when prices were high, the preceding harvest had been poor; when prices were low, the harvest had been bountiful. Using the so-called law of Gregory King, one might even calculate precisely the size of the harvest from the price at which grains sold.[27]

Such a simple, direct relationship between the local harvest and prices probably can be found in primitive, "closed" agricultural economies. Much of England, for example, in the early fourteenth century may have had such an economy. During the great famine from 1315–1322, harvest yields went down and grain prices went up inversely, as can be seen in Figure 1. And as prices rose, the poor starved. Such a model may also apply to the northwest of England during the famines of the late sixteenth and early seventeenth centuries. Unfortunately, for that region neither harvest yields nor prices are available, although concluding that it was a closed economy, at least with regard to grains, is not inconsistent with what is known about that poor and isolated area. In closed economies starvation was often a direct result of harvest failure which was the result of bad weather.[28]

The danger of fatal famines—brought by bad weather—was particularly intense in these economies after a long, sustained increase in population had reduced the per capita agricultural output to the subsistence level, even in good years. The population growth of the twelfth century had brought, by 1300, an uneasy equilibrium between population and resources in southern England; the population increase of the sixteenth century led to the same dangerous situation in the northwest of England. It is no accident that the great famine of 1315–1322 in the south and the famines of 1597 and 1623 in the northwest of England came more or less at the beginning of cold, unstable climatic periods.

27 W. G. Hoskins, "Harvest Fluctuations and English Economic History, 1480–1619," *Agricultural History Review*, XII (1964), 28–46; *idem*, "Harvest Fluctuations and English Economic History, 1620–1759," *ibid.*, XVI (1968), 15–31; B. H. Slicher van Bath (trans. Olive Ordish), *The Agrarian History of Western Europe A.D. 500–1850* (London, 1963), 118. Gregory King's law is described in *ibid.*, 118.

28 Ian Kershaw, "The Great Famine and Agrarian Crisis in England 1315–1322," *Past & Present*, 59 (1973), 3–50.

Fig. 1 Grain Prices and Yields

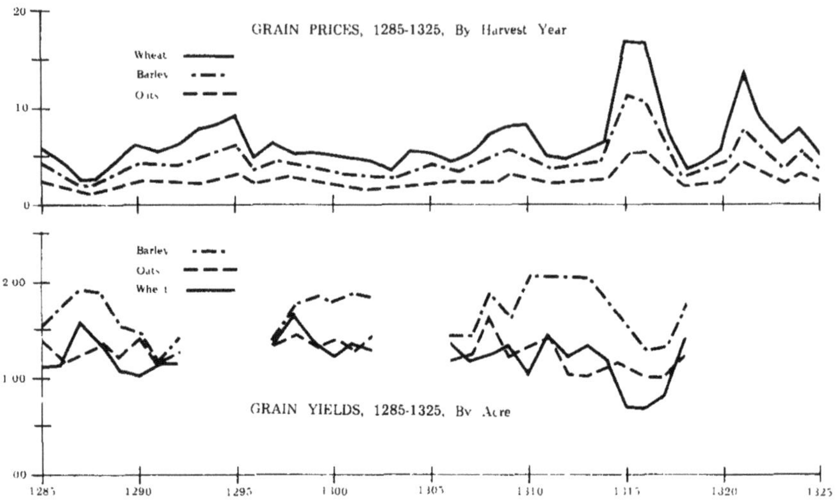

SOURCES: D. L. Farmer, "Some Grain Price Movements in Thirteenth-Century England," *Economic History Review*, X (1957), 212; J. Z. Titow, *Winchester Yields: A Study in Medieval Productivity* (Cambridge, 1972), 85–108

The relative warmth and stability of the weather in the thirteenth century may have favored settlement on marginal agricultural land and allowed populations to grow to a size too large to be sustained when the weather became less favorable. It is largely speculation, but the same process may have later taken place in the northwest.

During the relatively warm period from 1500 to 1550, the inhabitants may have cultivated upland areas which then could grow crops, but, with the coming of cooler weather, these lands were no longer productive. A drop in the average summer temperature of one degree centigrade is the equivalent of raising the level of the land surface 500 feet; in the northern English uplands such a temperature fall would have had an appreciable impact on the production of basic grains and pushed the population closer to subsistence. In addition, the cooling that began about 1550 would have increased the chances of successive harvest failures throughout the highlands of northwestern England.[29]

29 Emmanuel Le Roy Ladurie, *Histoire du climat depuis l'an mil* (Paris, 1967), 238; Wrigley, *Population and History* (London, 1969), 79; M. L. Parry, "The Significance of the Variability of Summer Warmth in Upland Britain," *Weather*, XXXI (1976), 212–217.

Climate, therefore, has been a factor of importance in the famines of the late sixteenth and early seventeenth centuries. But even in the northwest of England, climatic deterioration was only one of the many factors contributing to starvation. The region's cottage textile industry was in prolonged decline at the time of the famines and the cheap, low-quality cloths were also unusually sensitive to short-term falls in demand when food prices were high in other parts of England. The area was ideally suited to the raising of livestock, not arable crops, but livestock prices did not advance as rapidly as did grain prices after 1550. Thus the region was forced to grow its own grain rather than import what it needed, with considerable risk, as we have indicated.[30]

Southern England was not as closed economically nor as saddled with economic problems as was the northwest, even though the south had surplus people and many areas suffered from the occasional depressions in the clothing industry. Nothing is known about harvest yields in the late sixteenth or early seventeenth centuries and the impact of the weather on harvest and prices cannot be determined. Recent work, particularly that of Tits-Dieuaide on the southern Netherlands in the fifteenth century and Pfister on eighteenth century western Switzerland, has shown that the price mechanism for grains in developed, open economies was more complex than can be accounted for by any local weather, harvest, and price relationship. In any given year, the price of grain was influenced not only by the preceding local grain harvest (and presumably the local weather), but also by harvests in other areas, by the amount of grain in storage from earlier harvests, by speculation, and by governmental action.[31]

All of these factors affected English grain prices as well, at least in times of shortage. The English grain prices would have been dampened somewhat by the sale of cheaper grains from the Baltic in 1597 and 1638. In 1621, the harvest was very poor but prices did not rise because grains in storage from the bumper harvests of 1619 and 1620 were released on the market, making good the shortfall from that year's harvest. (Probably the northwest of England never enjoyed a surplus; its people may have

30 Appleby, *Famine*, passim.

31 M.-J. Tits-Dieuaide, *La formation des prix céréaliers en Brabant et en Flandre au XVe siècle* (Brussels, 1975), xi-xvii, 118–121; Pfister, *Agrarkonjunktur und witterungsverlauf im westlichen Schweizer Mittelland 1755–1797* (Bern, 1975), passim; *idem*, "Climate and Economy," 228.

depended on each year's harvest, without any carryover from one year to the next.) According to contemporaries, speculation in grains affected the price, as did the government's attempts to bring hoarded grain to the market.[32]

The ability to store grain, to buy supplies abroad, and to curb speculation did not free southern England from subsistence crises but it does seem to have reduced the threat of any single bad harvest, and only the prolonged nasty weather during the 1590s and 1630s had any serious demographic consequences. From about 1650—the exact date is in doubt because of poor parish registration from 1640 to 1660—England was free of subsistence problems until the late 1720s, when a purely local crisis occurred in a few Midland parishes. Harvest failures happened regularly after 1650, as they had before, but England managed to prevent harvest failures from turning into famine at the same time that France was becoming especially vulnerable to famines.[33]

The difference between the two countries, both of which lie in approximately the same climatic zone, suggests that famine was not so much a climatic problem as a result of economic, social, and governmental failures. The weather was much the same in the two countries; English wheat prices rose in the same years—1661–1662, 1693–1694, 1709–1710—that the French experienced famine, although the winter of 1709 may not have been quite as severe in England as in France.[34]

Why then was England able to solve the problem of feeding its people in the mid-seventeenth century—during the little ice age—when France was not? The question has been surprisingly little studied, but there are possible, rather conjectural, answers.

England was a more open economy than France, or at least than the interior provinces of France, where difficulties of transportation and impediments to the free movement of grain would have made the local people more directly dependent on their own harvests and less able to import outside food to alleviate local shortages. But the grain-growing plains of northern France, the

32 Peter Bowden, "Agricultural Prices, Farm Profits, and Rents," in Joan Thirsk (ed.), *The Agrarian History of England and Wales,* IV: *1500–1640* (Cambridge, 1967), 618; R. Lemon and M. A. E. Green (eds.), *Calendar of State Papers, Domestic, 1596–1597* (London, 1857–), 421; personal communication from Martin Ingram; Appleby, *Famine,* 142–144.

33 See note 2 above.

34 Hoskins, "Harvest Fluctuations, 1620–1759," 16–31; Lamb *Climate,* II, 466n.

Parisian basin, and the Loire Valley—where the famines were most lethal—were all open, with a well-developed grain trade and adequate transportation along the major rivers. Thus this is at best a partial explanation. (The same rainy weather that damaged the crops sometimes flooded the rivers and made grain shipments more difficult, but the English faced the same problem.)[35]

England had a better system of local parish relief than did rural France, which depended on voluntary charity. The English Poor Law of 1601—originally passed, hardly by coincidence, during the famine of 1597—provided for compulsory rates levied on the wealthier inhabitants of each parish to provide for their less-fortunate neighbors. Local parish relief reduced mortality in two important ways. First, if effective, it eliminated starvation in the countryside. Second, it cut down the flow into the cities of indigents seeking charity, thereby reducing the spread of disease. Unfortunately, however, we do not know whether English poor relief was effective or not. The example of a mini-crisis in certain Midland parishes in the late 1720s suggests that relief was not universally effective or could be overwhelmed if transportation costs were high, as they were into the Midlands.[36]

Nor did those in the middle income bracket in England have the oppressive tax burden that fell on poorer groups in France, particularly during the great wars at the end of Louis XIV's reign. It is difficult to weigh the contribution of taxation to the French famines but taxes would have pushed many rural smallholders nearer subsistence, leaving them with less of a cushion when the harvest failed. Heavy taxation also may have dried up a certain amount of rural money that otherwise might have gone to charity.

Finally, England enjoyed an agricultural revolution during the seventeenth century which France did not. In the early decades of the century, England imported grains in times of dearth; by the end of the century, she had become a regular exporter, even

35 See, for example, Michel Bricourt, Marcel Lachiver, and Julian Queruel, "La crise de subsistance des années 1740 dans le ressort du Parlement de Paris," *Annales de démographie historique* (1974), 320–321.

36 J.-P. Gutton, *La société et les pauvres* (Paris, 1971), 174; Cissie Fairchilds, *Poverty and Charity in Aix-en-Provence* (Baltimore, 1976), 107–108; Lebrun, *Les hommes et la mort,* 274; Goubert, "Le regime démographique français au temps de Louis XIV," in E. Labrousse et al., *Histoire économique et sociale de la France* (Paris, 1970), II, 43; B. Darivas, "Étude sur la crise économique de 1593–1597 en Angleterre et la loi des pauvres," *Revue d'histoire économique et sociale,* XXX (1952), 382–398.

during the 1690s and after the harvest failure of 1709. (Only in 1728 and 1729 did imports again exceed exports.) An increase in agricultural output by itself does not, however, explain England's victory over famine, particularly if the agricultural revolution came at the expense of English smallholders who were booted off the land, as Brenner has argued. Perhaps part of the new profits from agriculture went into poor relief. Was the smallholder first pushed off the land by his landlord and then supported by the landlord's charity? Possibly the agricultural revolution brought a better balance to crops in England and some protection against the weather, as I argue elsewhere.[37]

Far from experiencing any revolution, French agriculture underwent a cyclical decline in output during the latter years of Louis XIV's reign. Following a period of good yields from 1662 to 1688, output fell until about 1720. Such a decline would have pushed many of the poor closer to subsistence and may have eliminated any surplus that could otherwise have been stored even in relatively good years. With such a tenuous balance between people and resources, the harvest failures of 1693 and 1709 had particularly unpleasant results.[38]

The exact role that each of these factors played in contributing to famine in France or in enabling England to avoid famine is as yet unclear. But they represent known differences between the two societies and we have no evidence that the climate was basically different. With the evidence available, we can claim that the weather was crucial only where economic, social, and governmental protections were lacking.

After 1709, France did not suffer another major famine until 1795. In 1739–1740, however, crops failed over large parts of northern France and prices rose alarmingly to levels similar to those that obtained after the harvest failure of 1709. In this latter crisis, famine was averted through the vigorous action of the

37 Eric Kerridge, *The Agricultural Revolution* (New York, 1968), passim; A. H. John, "English Agricultural Improvements and Grain Exports, 1660–1765," in D. C. Coleman and John (eds.), *Trade, Government and Economy in Pre-Industrial England* (London, 1976), 45–67; Robert Brenner, "Agrarian Class Structure and Economic Development in Pre-Industrial Europe," *Past & Present*, 70 (1976), 30–75; Appleby, "Grain Prices and Subsistence Crises in England and France, 1590–1740," *Journal of Economic History*, XXXIX (1979), 865–887.

38 Le Roy Ladurie and J. Goy, "La dîme et le reste XIVe-XVIIIe siècle," *Revue historique*, CCLX (1978), 123–142.

Parlement of Paris, the urban authorities, and the royal government. Grains were imported into the affected areas, the rich were taxed to support the poor (how effectively is not known), workshops were established, brewing was prohibited, money was given to the indigent, and other measures too numerous to list were enforced. Starvation was prevented, although the crisis years were marked by much illness.[39]

The example of 1740 shows that the French authorities were capable of preventing famine, even in years of dreadful weather and widespread harvest failure. Again in 1816, following another dismal harvest, the French authorities successfully averted famine. In short, governmental action could—and on occasion did—prevent famine after even the worst climatic shocks.[40]

I have argued here that England and then France adapted to the occasional spell of bad weather that reduced the harvest. At different times both countries overcame the problem of famine through increased governmental protection of the poor, through improvements in food distribution, and through agricultural advances that were made within the existing technology and with crops that had long been available. The crucial variable in the elimination of famine was not the weather but the ability to adapt to the weather. Certainly a successful adaptation was more difficult in areas of marginal cultivation—such as Scandinavia, Scotland, and the mountainous regions throughout Europe—where climate played a larger role in fluctuating crop yields. But in the temperate regions of England and France human responses to the climate were more important than the climate itself both in causing famine and in eliminating it.

39 Bricourt et al., "La crise de subsistance des années 1740," 297–302, 315–320.
40 Post, *Last Great Subsistence Crisis,* 53–67.

Christian Pfister

The Little Ice Age: Thermal and Wetness Indices for Central Europe

The Little Ice Age: Thermal and Wetness Indices for Central Europe The kinds of evidence used to reconstruct past weather and climate can be divided into two classes: for those generated through natural manifestations of climatic change and analyzed by scientific disciplines, the term *field data* is appropriate; for the body of man-made climatic data in the form of written and illustrated documents buried in archives, libraries, and museums we use the term *documentary data*. Only recently has it been shown that documentary sources of information about past climates are not equally reliable. Much material which purports to record historical events is gravely misleading. Almost all compilations of weather descriptions, printings, and manuscripts include events which are non-contemporary. Often observations have been copied (or miscopied) from other sources, sometimes even without giving a reference.[1]

We distinguish *two groups of documentary data* in Table 1. The first group includes instrumental *measurements* and several types of non-instrumental *observations* for which individual *weather factors* (temperature, precipitation, wind, etc.) are specified. We include among the measurements only those which are found in archives and libraries. Most of them were made by private individuals before networks were created by national weather services.

The second group, *documentary proxy data*, summarizes a variety of information which reflects the combined effect of several weather factors, during a period of several months. Like the field data of the scientist, documentary proxy data can be calibrated with instrumental measurements and used to estimate specific meteorological variables.

Christian Pfister is Research Fellow of the Swiss National Science Foundation.

Acknowledgments are due to Bruno Messerli and John D. Post for reading the manuscript and making helpful suggestions and corrections. Preparatory work for this article was made possible by a post-doctoral fellowship from the Swiss National Science Foundation.

1 Martin J. Ingram and David J. Underhill, "Historical Climatology," *Nature*, CCLXXVI (1978), 329–334; *idem*, "The Use of Documentary Sources for the Study of Past Climate," paper given at the Conference on Climate and History (Univ. of East Anglia, 1979), 59–91.

Table 1 A Taxonomy of Data Used to Reconstruct Past Weather and Climate

ORIGIN / INDIVIDUAL WEATHER FACTORS	FIELD DATA	DOCUMENTARY DATA (archives, libraries, museums)
specified		*Measurements, observations* — Chronicles, annals, etc. — Non instrumental diaries — Measurements taken prior to the creation of national networks
nonspecified	— Pollen — Tree rings — Isotopes — Moraines — Varves etc.	*Documentary proxy data* — Phenological observations — Wine and grain harvest dates from administrative sources (paraphenological data) — Quantity and quality of wine produced — Illustrated sources of glaciers

The first part of this article presents three types of documentary proxy data (phenological observations, grain harvest dates, and vine yields) and shows how they are interpreted, calibrated, and cross-dated to non-instrumental observations. In the second part a quantification of the whole body of observations—specified and unspecified—is provided in the form of an index for both temperature and precipitation with a time resolution of a month. It is then demonstrated to what extent such a great body of highly detailed information may modify the periodization of the little ice age in Central Europe.[2]

A data bank of documentary weather evidence (CLIMHIST) has been compiled following a systematic search for evidence in the major libraries and archives of Switzerland. The search revealed more than 27,000 records, mainly from places on the plateau. The records include monthly means of the Basel temperatures (from

2 Phenology is the science which relates periodic biological phenomena (crop growth, migration of animals, etc.) to weather and climate. The stages of development of a certain species (burgeoning, flowering, ripening, etc.) are called phenophases.

1755); several series of rainfall measurements (from 1708); summaries of more than 70,000 daily observations from weather diaries; some 3,000 observations of snow-cover and snowfalls on the alpine pastures; almost 3,000 phenological observations; and reports on the conditions of crops and cattle, the quality of harvests, and the occurrence of diseases.

A numerical code assigned to the descriptive observations facilitated the management of the data by computer. The printout has the form of a weather chronology in readable form. It contains information about the weather and its impact upon the hydro-, bio-, and anthroposphere from 1525 to 1825 with time intervals from ten days up to entire seasons.[3]

CONVERTING DOCUMENTARY PHENOLOGICAL MATERIAL INTO CLIMATIC DATA Occasional phenological observations of crops or trees are contained in many chronicles and annals. They have never been used as climatic evidence, probably because they were too erratic and could not easily be interpreted. Given that the fluctuations of phenophases are in good agreement over distances of several hundred kilometers, the Swiss evidence, which is presented below, may also be conclusive for the climate in great parts of Central Europe.[4]

In most cases single phenological observations describe the growth pattern of outstanding years. Plants are known to be living instruments which show in their growth response the composite effect of temperature, rainfall, sunshine, and radiation. In most cases temperature is the dominant variable. Hence a reference to the growth calendar of plants was the most objective way

3 Pfister, "Klimageschichte der Schweiz 1525–1825," in preparation. The CLIMHIST data bank (compiled by Pfister) is stored on disks and could easily be included in a larger international data bank. *Idem,* "The Reconstruction of Past Climate: The Example of the Swiss Historical Weather Documentation," paper given at the Conference on Climate and History (Univ. of East Anglia, 1979), 134.

4 Helmut Lieth (ed.), *Phenology and Seasonality Modelling* (New York, 1974), preface; John A. Kington, "An Application of Phenological Data to Historical Climatology," *Weather,* XXIX (1974), 320–328; Fritz Schnelle, "Temperaturverhältnisse und Pflanzenentwicklung in der Zeit von 1731 bis 1740 in Mittel-und Westeuropa," *Meteorologische Rundschau,* XI (1959), 58–63; *idem,* "Hundert Jahre phänologische Beobachtungen im Rhein-Main Gebiet 1841–1939, 1867–1947. Ein Beitrag zur Klimageschichte des Rhein-Main Gebiets," *Meteorologische Rundschau,* II (1950), 150–156; Max Bider, "Untersuchungen an einer 67-jährigen Reihe von Beobachtungen der Kirschblüte bei Liestal (Basel-Landschaft)," *Wetter und Leben,* XII (1960), 36–50.

of documenting the coldness or the warmth of a particular season before the thermometer was invented. It implied that the reader, who was supposed to be familiar with agriculture, could compare it to the "normal" pace of vegetative growth. Whether the historian of climate can decode this material and convert it into valid climatic data depends entirely on the quality of the evidence which can be found. Two conditions must be met in order to be able to decode the phenological information.

First, phenological series of sufficient length (fifteen to twenty years at least) are required for a period which is fully documented by meteorological measurement (temperature, precipitation, and duration of sunshine). Such a series allows for quantification of the relationship between the growth pattern of a certain species and the environmental factors. Although modern phenology seeks to "explain" biological processes in terms of environmental parameters, the historian of climate, who uses phenology as a substitute for meteorological measurement, has to take temperature as the dependent variable and the phenological observations as the independent variable. The study must also include ecological theory and may not be restricted to statistical manipulations.[5]

In many countries a considerable body of observations has already been collected. In Switzerland in particular a few series have survived from the eighteenth century but, surprisingly, more recent observations are difficult to find because regular network observations were not begun until three decades ago. The only record which is both sufficient in length and also covered by a full set of meteorological measurements has been maintained by the observers of three meteorological stations in the Canton of Schaffhausen (northern Switzerland) from the end of the nineteenth century until about 1950. These records were used to

5 Lieth, *Phenology*, 3–19. In particular, attitude and exposure of the place on which the plant grows have to be taken into account. Norbert Becker, "Phänologische Beobachtungenan Reben und ihre praktische Anwendung zur Gütekartierung von Weinbergslagen," *Weinwissenschaft*, XXIV (1969), 142, has found that for the vine flower an increase in altitude of 10 m corresponds to a delay of 0.36 days, if the other factors are held constant. Richard Volz, "Phänologische Karte von Frühling, Sommer and Herbst als Hilfsmittel für eine Klimatische Gliederung des Kantons Bern," *Jahrbuch der Geographischen Gesellschaft Bern*, LII (1975/6), 46–52, has found that for apple bloom the mean delay between the earliest appearance (south) and the latest appearance (east) is 4.5 days. For the wheat harvest no clear pattern has emerged.

calibrate the body of historical evidence contained in the CLIMHIST data bank (see Table 2).[6]

Second, because phenological observations are only conclusive in the form of deviations from a mean, we need to determine whether the means have changed from those of the past and whether such changes were the result of the changing climate or the introduction of new varieties. The results are discussed in Figure 1.

When we compare recent and historical means it is surprising to find that they differ only by one or two days, which does not affect the kind of rough estimates which we have given.[7] Table 2 displays the calibration of phenophases at the most conclusive meteorological statistics.

THE EXTREMES OF INDIVIDUAL PHENOPHASES

Sweet cherry flower. Based upon the longest known phenological record (from the ninth to the nineteenth century) which relates to a variety of cherry tree in Japan, Arakawa has demonstrated the close correlation between spring temperatures and flowering. We can deduce from Table 2 that his conclusion agrees with the result obtained from the analysis of the Unter-Hallau series. Temperatures in March seem to be far more important than in April.[8]

If the cherry flower were advanced by more than two weeks, we may conclude that either February and March had been much

6 Richard J. Hopp, "Plant Phenology Observation Networks," in Lieth, *Phenology*, 25–43, for a survey of the historical series and an exhaustive bibliography. From 1611 to 1644 the beginning of the rye harvest is reported in the "Chronicle of Abraham Künzli," unpub. ms., Stadtbibliothek, Winterthur Ms. Q 72. For 1721 to 1738, Pfister, "Zum Klima des Raumes Zürich im späten 17. und frühen 18. Jahrhundert," *Vierteljahrsschrift der Naturforschenden Gesellschaft Zürich*, CXXII (1977), 447–471. For 1760 to 1802, *idem*, *Agrarkonjunktur und Witterungsverlauf im westlichen Schweizer Mittelland* (Bern, 1975), 73–78. For the nineteenth and twentieth centuries, *idem*, "Local Phenological Time Series from the Canton of Schaffhausen (Switzerland) and their Application for the Interpretation of Historical Records," unpub. ms.

7 An exception is the events of the Winterthur series (1721–1738), which are considerably more advanced than the recent means, and may be attributed to the high frequency of early springs in the 1720s.

8 H. Arakawa, "Twelve Centuries of Blooming Dates of the Cherry Blossoms at the City of Kyoto and its own Vicinity," *Geofisica pura e applicata*, XXX (1955), 36–50. The results for Unter-Hallau agree well with those of Bider, "Untersuchungen," 9–10.

Fig. 1 Extremes of Vegetative Development, 1525–1825 in the Swiss Lowlands (400–500 m).[a]

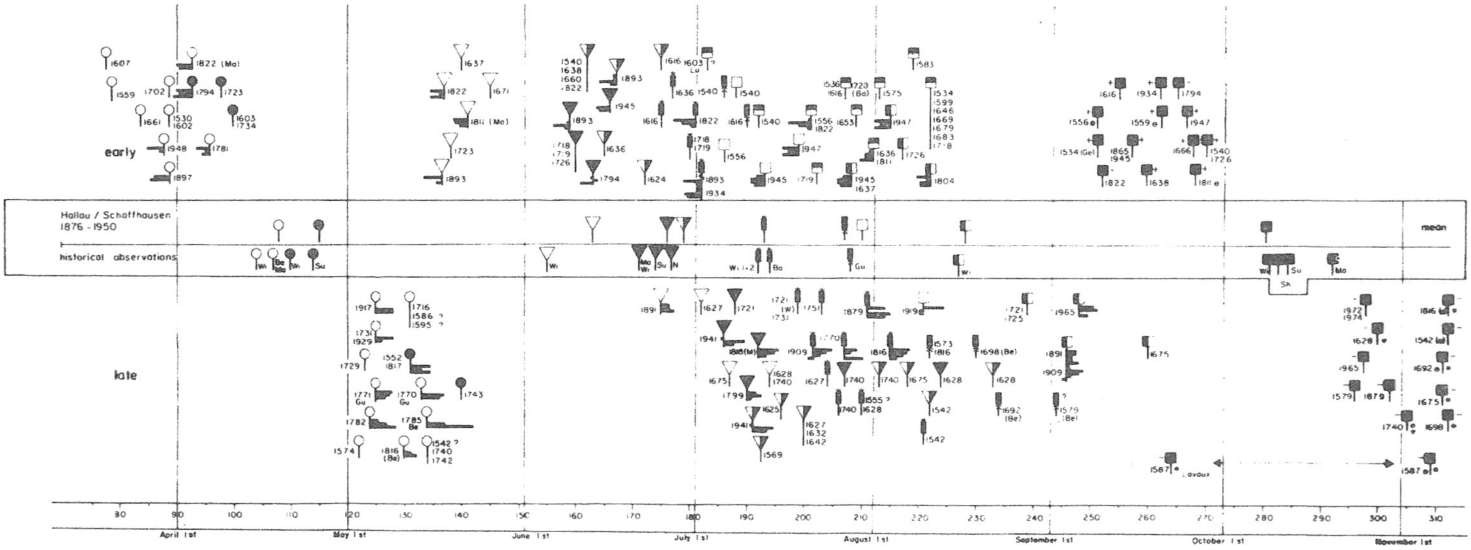

Legend for Fig. 1

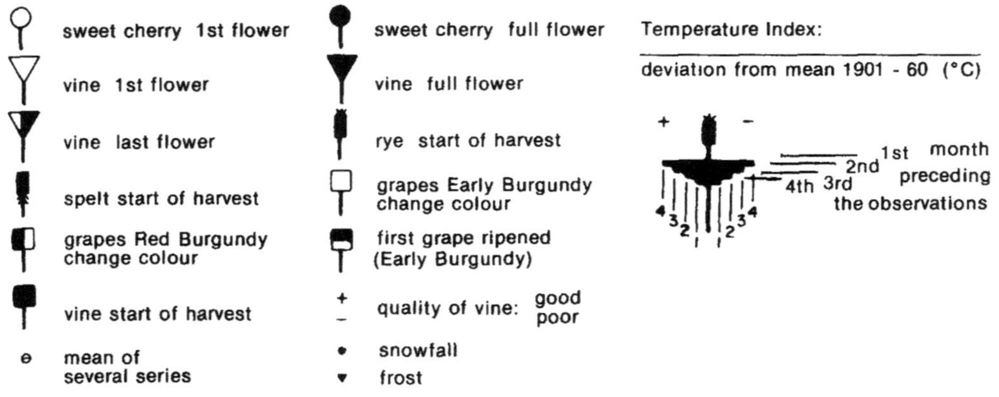

Historical series and single observations outside the altitude range 400 - 500 m

Ba	Basel	256m	
Ba	Batterkinden BE	473m	1718 - 70
Be	Bern	540m	1780 - 1827
Ge	Geneve	375m	
Gu	Gurzelen BE	591m	1766 - 84
Ma	Maschlins GR	534m	1792 - 1818
N	Neuchatel	447m	1701 - 50
Sh	Schaffhausen	403m	1616 - 70
Su	Sutz BE	463m	1785 - 1803
Wi 1	Winterthur	443m	1611 - 44
Wi 2	Winterthur	443m	1721 - 38

a Figure 1 provides a survey of means, maxima, and minima of selected phenological events, both modern and historical. The dates have been converted to days of the year and are shown on the x-axis. Symbols are used to express the different phenophases. Within the frame, which divides the graph into an upper and a lower part, the means are represented; the recent ones have been taken from the Schaffhausen series, and the historical ones from various records, which are listed in the legend The earliest single events, which are contained in the historical and the recent record, are plotted above the frame, whereas the latest ones are plotted below.

Because the date of phenological events is also affected by altitude, the documentation refers mainly to phenophases observed at altitudes from 400 to 500 m (events outside this range are marked with letters in parentheses). Differences in exposure were not considered. Whenever an event occurred after the beginning of the thermometrical measurement in Basel (1755), horizontal bars are drawn, which represent the deviations of temperature from the 1901 to 1960 average (positive to the left, negative to the right) during the preceding months. Thus observations which are calibrated with measured temperatures can easily be compared with data from the pre-meteorological period.

SOURCE. CLIMHIST data bank.

Table 2 Selected Phenophases at the Meteorological Station of Unter-Hallau (alt. 430 m) in Northern Switzerland and Their Response to Temperature

PHENOPHASE	N	MEAN DATE	DETERMINED BY TEMPERATURES IN				
				PARTIAL R^2		PARTIAL R^2	TOTAL R^2
Sweet cherry flower	54	April 16 (day 106)	March	48	April	12	60
Vine first flower	61	June 9 (day 160)	May	59			59
Rye start of harvest	45	July 12 (day 193)	May	18	June	29	47
Wine harvest (variety Red Burgundy)	70	Oct 7 (day 280)	April to June	54	July to Sept	9	63

N = number of observations
R^2 = proportion of total variance explained by the mean temperature of the listed months.
SOURCE: Pfister, "Local Phenological Time Series."

above average (1822, 1897, 1794) or that January had been extraordinarily warm and February and March had been somewhat above average (1948). For 1607, when the cherry blossom appeared four weeks too early, Renward Cysat reported that there was no winter at all. The ground was never frozen or covered with snow, the sun shone most of the time, the vegetation did not come to a standstill, and people were wearing summer clothes. We may estimate that during this "year without a winter" January and February may both have been as warm as March, on average, which has never been the case from 1755 to the present. The cherry blossomed nearly as early in 1602 and 1603. However, Cysat reports considerable delays in the spring vegetation in 1600, 1601, and 1608, which suggests that outstanding weather patterns prevailed in spring during that first decade of the seventeenth century.[9]

A delay of flowering by three weeks or more suggests that the March-April period may have been at least 5° too cold. An extreme case was March 1785, which was 8° below the mean. The delays in 1716, 1740, 1770, and 1817 can be attributed more to an unusually cold April.[10]

9 Renward Cysat (ed. Joseph Schmid), *Collectanea pro Chronica Lucernensi et Helvetiae* (Luzern, 1969), I, Pt. 2, 907, 908, 945.

10 Bider, Max Schüepp, and Hans von Rudloff, "Die Reduktion der 200 jährigen Basler Temperaturreihe," *Archiv für Meteorologie, Geophysik und Bioklimatologie,* IX (1959), 360–412. Examples for 1716 and 1740 are contained in the CLIMHIST data bank.

Vine flower. We should consider only the observations which were carried out in an open vineyard. Plants which are sheltered by the wall of a house will flower considerably earlier. Differences in varieties, however, can be neglected. Although the first flower is advanced or delayed mainly according to temperature in May (see Table 2), an early flower may also follow a very warm April, as was the case in 1811 and 1893 and, according to descriptive evidence, in 1723. The times of the full bloom and the last flower vary with temperatures in both May and June. In all those years, for which an early flowering was reported (e.g. 1636–1638, 1660, and 1718), the wine harvests throughout Central and Western Europe were also very advanced. Thus if the evidence suggests that the vine flowering ended in the first half of June, the advance of the wine harvest should be credited mainly to the warmth of spring.[11]

Extreme delays (1542, 1627, 1628, 1632, 1642, 1675, and 1740) were much more frequent and much more pronounced than extreme advances.

Start of cereal harvest. Today the decision to begin the harvest is made by an individual and, therefore, is affected by economic and social factors as well as purely climatic ones. But, while the three field system was in use (i.e. until the early nineteenth century) agreement was reached jointly by the farmers of a village. In the eastern part of Switzerland, where spelt and rye were grown, the latter was harvested about two weeks earlier on averge. The maturity of rye is controlled by temperatures in June and, to a lesser extent, by those in May (see Table 2). Again the very early rye harvests (1616, 1636, 1718, 1719, and 1822) as well as the earliest spelt harvest (1540) were clearly connected to very early wine harvests.

11 Hermann Trenkle, "Die Verwendung phänologisch-klimatologischer Beobachtungen bei der Gütebewertung von Weinbergslagen," *Weinwissenschaft,* XXIV (1969), 327–338. The significance of temperatures in May for an early flowering is also emphasized by Ernst Peyer and Werner Koblet, "Der Einfluss der Temperatur und der Sonnenstunden auf den Blütezeitpunkt der Reben," *Schweiz. Zeitschrift für Obst-und Weinbau,* CXII (1966), 250–255; W. Hofäcker, *Einfluss von Umweltfaktoren auf Ertrag und Mostqualität der Rebe* (Hohenheim, 1974); Becker, "Oekologische Kriterien für die Abgrenzung des Rebgeländes in den nördlichen Weinbaugebieten," *Weinwissenschaft,* XXXII (1977), 77–102. Emmanuel Le Roy Ladurie and Micheline Baulant, "Grape Harvests from the Fifteenth through the Nineteenth Centuries," in this issue, provides a main series composed from 102 local series of wine harvest dates.

Delayed harvests of rye and spelt reflect temperatures from May to July. Those in 1816 and 1879 were affected by extremely low temperatures, which were the lowest ever recorded in Basel during that period. Again the most deferred cereal harvests agree almost completely with the latest vintages of the little ice age (1542, 1879, 1740, 1816, 1698, 1555, 1628, and 1573).[12]

Maturation of grapes and date of the vintage. Chronicles from vine-growing regions frequently reported the date on which the coloration of grapes was observed for the first time or gave the date on which the first grape reached maturity. In interpreting this evidence we have to differentiate between early and late varieties.[13]

Wine harvest dates have been taken as good proxies for mean temperatures during the period between April and September, and may give a general idea of the spring-summer temperatures for the last 500 years. If, however, we want to crossdate wine harvest dates with more detailed qualitative information and use them to estimate the temperature patterns of a particular season or month, a more sophisticated model is needed. A multivariate statistical analysis of the Unter-Hallau series has revealed that wine harvest dates are more closely correlated with temperatures from April to June than with those from July to September (see Table 2).[14]

Several authors have shown that the growth of the grapes after fructification includes a phase of standstill of up to three weeks in duration which, in locations with favorable climatic conditions, occurs in August; this could explain the non significant correlation for this month. The low impact of temperatures in April and September may reflect the fact that growth is not

12 *Ibid.*

13 According to Figure 1 the coloration of the Early Burgundy grape, which was very widespread in the past, sets in fifteen days earlier than that of the Red Burgundy (Pinot Noir) grape. According to Trenkle, "Weinbergslagen," the mean date for the Müller Thurgau grape is Oct. 6, and for the Riesling is Oct. 29, to mention the extremes.

14 Marcel Garnier, "Contribution de la phénologie à l'étude des variations climatiques," *La Meteorologie,* XL (1955), 291–300; Le Roy Ladurie (trans. Barbara Bray), *Times of Feast, Times of Famine: A History of Climate Since the Year 1000* (Garden City, 1971), 50; Le Roy Ladurie and Baulant, "Grape Harvests." Correlations with the monthly mean temperatures: June (−.59), significance 0.001; May (−.54), significance 0.001; July (−.34), significance 0.03; September, April, and August were not significant.

activated as long as temperatures are below 10°. The duration of daylight, which plays an important role, could also account for the importance of temperatures in early summer.[15]

It is not surprising that the very late wine harvest dates show almost no temporal dispersion. If the grapes failed to reach maturity, which was not unusual in the Swiss vineyards during the little ice age, the harvest decision was triggered by early frosts and snowfalls. Thus these dates are not really conclusive for the rank order of the coldest spring-summer seasons.

Instead of focusing upon the analysis of certain phenophases, the whole pattern of phenophases from spring to autumn may be examined. From Figure 1 two types of phenophases can clearly be distinguished.[16]

1. *All phenophases advanced*
   - Documented with temperature measurements: 1781, 1794, 1811, 1822, 1893, 1934, and 1945.
   - Cases from the pre-meteorological period: 1540, 1559, 1599 (evidence not entirely reliable), 1603, 1604, 1637, and 1719.[17]
   - Temperature pattern: (derived from the Basel series) two or all of the spring months and June above the 1901–1960 average; July, average or above.
2. *All phenophases delayed*
   - Documented with temperature measurements: 1770, 1816, 1817, 1879, 1891, and 1909.
   - Cases from the pre-meteorological period: 1542, 1573, 1627, 1628, 1716, and 1740.
   - Temperature pattern derived from the Basel series: at least four months from March to July below the 1901–1960 average.

15 Peyer and Koblet, "Blütezeitpunkt"; G. Alleweldt, "Der Einfluss des Klimas auf Ertrag und Mostqualität der Reben," *Rebe und Wein,* XX (1967), 312–317; Pierre Basler, "Beeinflussung von Leistungsmerkmalen der Weinrebe (Vitis vinifera L.) in der Ostschweiz durch Klimafaktoren und Erträge sowie Versuch einer Qualitätsprognose, *Weinwissenschaft,* forthcoming.

16 The list is not exhaustive. The evidence not represented in Figure 1 is contained in the CLIMHIST documentation.

17 In 1540 the vine was possibly retarded by extreme drought and high temperatures in July and August, as happened in 1947: *Bericht über den Weinbau des Kantons Schaffhausen* (Schaffhausen, 1947), 6; Becker, "Oekologische Kriterien," 89.

Major changes in the intervals between phenophases may also be indicative of a significant deviation from the mean temperature. A shortening always indicates that temperatures during the interval have been above average. In 1934, when May was almost 2° above the mean, the interval between the first sweet cherry flower and the first vine flower (fifty-four days on average) was only twenty-eight days. Similar cases (twenty-nine days in 1726 and thirty-two days in 1731) are supported with descriptive evidence. A drastic shortening of the interval between the first vine flower and the start of the rye harvest, about two weeks compared with thirty-three days on average, occurred in 1616. The rye reached maturity even six days earlier than in 1822, which was the hottest June since 1755 (4° above the 1901–1960 average). This early date suggests that the heatwave in June 1616, which is impressively described in the sources, was the most severe since at least 1525.

Major delays between phenophases indicate that temperatures during the interval were below average. In 1916, when the mean duration of vine flowering in the vineyards of the Canton of Zurich was twenty-seven days instead of nineteen days, on average, June was 3° below the 1901–1960 mean. In 1628 and 1740 the first vine flowers in the vineyards of Schaffhausen appeared around July 10. But, although the duration of flowering was average in 1740, which suggests a July temperature around the mean, it was extended some thirty-five days in 1628 (see Fig. 1). This late flowering points to an extreme cold spell, which is consistent with the high number of snowfalls on the alpine pastures reported from the same summer.[18]

RYE HARVEST DATES AND THEIR CLIMATOLOGICAL SIGNIFICANCE

It has been shown that the phenological observations of the chroniclers and early amateurs can be used as a yardstick to measure the thermal deficit or excess of the most extreme seasons of the little ice age. But even if we were to compile many documents of this type, we could never hope to bring together a continuous and homogeneous record.

However, the series of wine harvest dates collected by Le

18 Bider et al., "Basler Temperaturreihe"; CLIMHIST documentation.

Roy Ladurie is a model. His series was obtained from purely administrative documents, which were kept regularly year after year. Le Roy Ladurie made a similar attempt with cereal harvest dates for the south of France, but discarded this evidence, perhaps because the series was too rough and short.[19]

I have discovered in several Swiss archives a new type of climatic evidence which can be obtained by extracting tithe figures from county accounts. With these figures it is possible to draw up a long, continuous, and homogeneous series of valid proxies for grain harvest dates. Such a series enables us to make temperature estimates for the May-June period and to support our estimates with descriptions of particularly hot and cold spells from weather diaries.

Tithes paid in kind are known to be a fair guide to the size of grain harvests. They were influenced by several variables, notably the percentage tithed of the harvest, the acreage under cultivation, and the yield per acre. Yields per acre are known to be a function of the quantity and quality of labor input, the amount of fertilizer applied, and the meteorological conditions. But as Slicher van Bath has cautioned, the relationship between plant growth and weather patterns is more complicated than is assumed; in the case of wheat and winter grains in the temperate zone, the weather patterns affecting growth fall into eight different phases and extend over a twelve-month period.[20]

An element of the tithe accounts that has turned out to be sensitive to climate is the date of the tithe auction. Before the early nineteenth century the right to collect tithes from certain areas was sold by auction to tithe-farmers, who carried out the work of tithe collection for provincial governors and were entitled to keep a small proportion of the revenue for themselves.[21]

19 Le Roy Ladurie, *History of Climate*, 50, 271.

20 Francois Jeanneret and Philippe Vautier, *Kartierung der Klimaeignung für die Landwirtschaft in der Schweiz* (Bern, 1977); Pfister, *Agrarkonjunktur*, 111–121, for the past; Bernhard Stauffer and Alfred Lüthi, "Wirtschaftsgeschichtliche Quellen im Dienste der Klimaforschung," *Geographica Helvetica*, XXX (1975), 49–56, provided a significant correlation between a tithe curve from the Canton of Argovia (northern Switzerland) and the changing ratio of oxygen in an ice core from Camp Century, Greenland. B. H. Slicher van Bath, "Agriculture in the Vital Revolution," *Cambridge Economic History of Europe* (Cambridge, 1977), V, 42–132.

21 Pfister, "Climate and Economy in Eighteenth Century Switzerland," *Journal of Interdisciplinary History*, IX (1978), 223–243.

In order to assess the size of the harvest, the fields were inspected by peasants hired by the governor some days prior to the auction, which in turn preceded the beginning of the harvest. In several tithe accounts the date of the auction appears regularly every year, whereas in others it is missing. It can be shown that the dates of inspection and auction were carefully chosen according to the ripeness of the grain. "The tithes in the mountains [at altitudes of 600 to 800 m] have not been inspected yet because the fields are still far from maturity and yields cannot be assessed properly at this time," wrote the governor of the County of Bipp to the Council of Bern on August 8, 1770. But it was risky to delay the auction until the grain was overripe, because the kernels could drop out of the ears, as was reported in 1806. Often the tithe districts, tributary to a corporation or a governor, were situated at different altitudes. Thus appropriate auction dates, up to five in some counties, had to be fixed for each altitude.[22]

The mean dates of the series are essentially a function of altitude. The higher a set of districts was situated, the later the auction took place on average. The mean delay was 4.6 days per 100 m, which corresponds exactly to the figure obtained by Volz from similar modern data.[23]

Tithe auction dates can therefore be used as substitutes for phenological observations in the same way as wine harvest dates. A variety of non-climatic factors must also be taken into account, in particular the time constraints of the governor or his delegate.

Because no dated tithe auction records have survived from the sixteenth century and because tithes were gradually abolished in the nineteenth century in return for a compensation, tithe auction dates cover only a little more than 200 years (1611–1825). The forty-two series which could be brought together varied in length and were in most cases highly correlated.

Those records of sufficient length were aggregated into a main series. Within the interval for which there were thermometrical measurements the residuals were compared with the May and June deviations of the temperatures in May and June from the 1901–1960 average. The timing of the tithe auction allowed an estimation of temperature deviations in the May–June period

22 *Idem*, "Getreide-Erntebeginn und Frühsommertemperaturen im schweizerischen Mittelland seit dem frühen 17. Jahrhundert," *Geographica Helvetica*, XXXIV (1979), 23–35.
23 Volz, "Phänologische Karte," 48.

with a standard error of 0.6°C. An advance or a delay of the auction date by seven days corresponded roughly to a deviation of 1° from the 1901–1960 average in May and June. Cross-dating with descriptive evidence suggests that the proxy underestimates the size of the anomaly.[24]

Tithe auction dates were also highly correlated with the main series of wine harvest dates from Western and Central Europe.[25] Figure 2 compares the mean residuals of tithe auction dates with temperatures in May–June in central England. In general the fluctuations of the tithe curve and the temperature curve in central England are in good agreement.[26] The high frequency of extreme years at the beginning of the series is striking, notably during the intervals 1611–1617 and 1626–1638.

Comparisons with the present, which can be made from two phenological series, are conclusive. Whereas the mean of the early series on the rye harvest from Winterthur (1611–1644) is exactly the same as in the Unter-Hallau series (1888–1950) (see Fig. 1), the standard deviation is nearly twice as large. This reflects an enhanced variability of temperature from spell to spell and from year to year, which, according to Lamb, is characteristic of regimes with frequent blocking, or meridional (north–south) circulation patterns in middle latitudes. As mentioned above, springs and winters from 1600 to 1610 were frequently affected by similar weather patterns.[27]

Two major episodes in the climatic history of the seventeenth century, the warm period from 1676 to 1686 and the cold years from 1687 to the turn of the century, are clearly shown in the curve in Figure 2. From the interpretation of the tithe curve and the body of descriptive evidence it may be derived that the sharp contrast between the two periods operated mainly through a

24 The model is as follows: $Y' = 0.056 - (0.13676 * TR)$ where Y' is the estimated deviation of the temperatures in Basel in May-June from the 1901–1960 average, and TR the residual of the aggregated series of tithe auction dates.

$R^2 = .64$ standard error: .6°

Harold Fritts et al., "Past Climate Reconstructed from Tree Rings," in this issue, have observed the same feature in tree-ring structure.

25 Coefficient of correlation r = .7 for 209 paired observations. Significance 0.0001.

26 Coefficient of correlation r = .68 for 95 paired observations. Significance 0.001. Major deviations are revealed in the early 1660s, where the May-June period in Switzerland was warmer than in England.

27 Hubert H. Lamb, *Climate: Present, Past and Future* (London, 1977), II, 465–466.

Fig. 2 Comparison of the Tithe Auction Dates in the Swiss Plateau and Central England Temperatures in Early Summer.

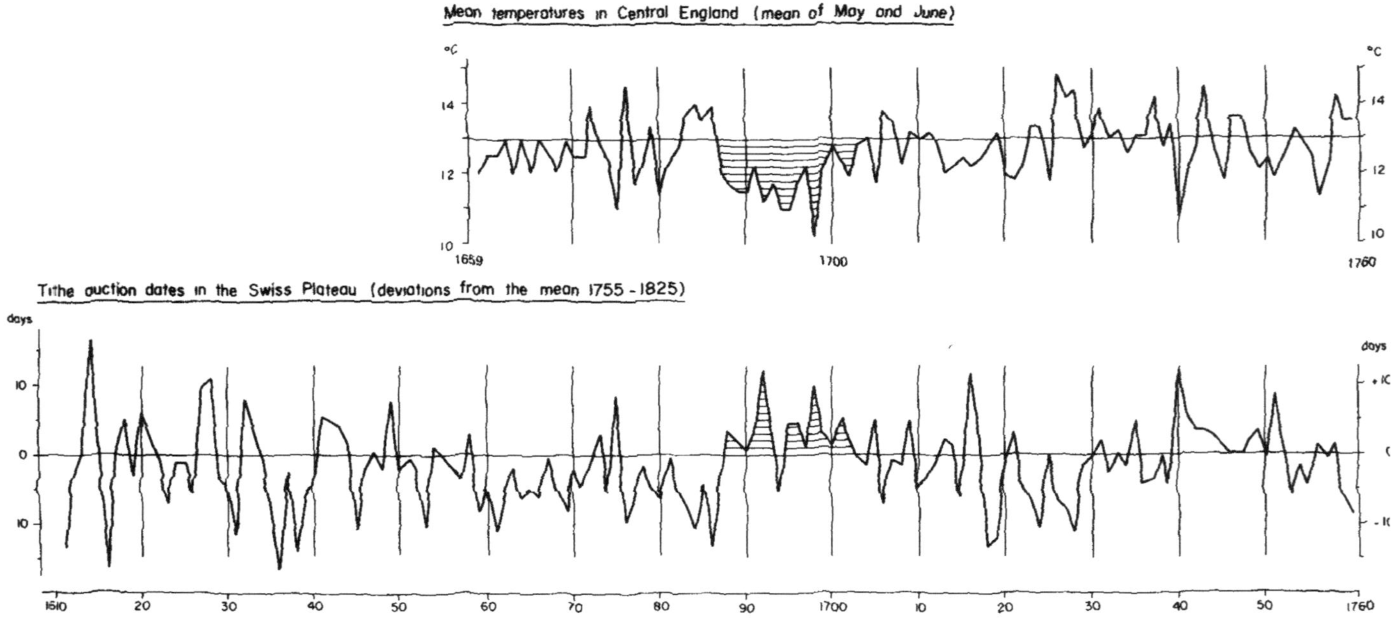

SOURCES: Temperature: Gordon Manley, "Central England Temperatures," *Quarterly Journal of the Royal Meteorological Society*, C (1974), Table 1. Tithe auction dates: Pfister; "Getreide-Erntebeginn und Frühsommertemperaturen," *Geographica Helvetica*, XXXIV (1979), Graph 4 (revised).

change in the prevailing weather in May. This month was repeatedly sunny, warm, and dry from 1676 to 1686. Then suddenly it became dull, chilly, and wet for more than a decade. These cold spells did not always extend over all of the summer months, as we might conclude from the prevailing interpretation of the wine harvest dates. Similarly temperatures in May (and sometimes June) retarded the vegetation during the 1740s. In both cases the interpretation of the retreats and advances of the two Grindelwald glaciers, by far the best documented in the historical past, pose contradictions as long as the wine harvest dates are taken as indicators of the quality of the summer months.[28]

FLUCTUATIONS OF VINE YIELDS AND MIDSUMMER TEMPERATURES

As an indicator of climate the grapevine has three major advantages:

1. The plant remains the same for twenty to fifty years. No annual planting is required.

2. The entire length of the growing season from March/April to October is needed to bring the grapes to maturity.

3. Harvest date, yield per acre and wine quality can be used as climatic proxy evidence for three different periods of the growing season: late spring/early summer, midsummer, and late summer/early autumn.

In comparison with the date of the harvest quantity and quality of wine have not often been utilized for the reconstruction of past climates, probably because reliable data from the preinstrumental period are difficult to find and because their interpretation is controversial.

Wine quality. Weger and Rima have found a cyclical quality in a bisecular series of wine yields from the estate of Johannisberg in the Rineland, which could be related to the fluctuations of sunspots. Wright has shown that the good and bad wine years in Luxemburg from the seventeenth century varied according to the warmth of summers in central England. Further back in time is the well-known wine chronicle for southern Germany by

28 Bruno Messerli et al., "Die Schwankungen des Unteren Grindelwaldgletschers seit dem Mittelalter. Ein interdisziplinärer Beitrag zur Klimageschichte," *Zeitschrift für Gletscherkunde und Glazialgeologie,* XI (1975), 12–50, 78.

Müller, which lists contemporary opinions on the size and the quality of the harvest from 1 A.D. to 1950.[29]

Wine quality is generally a function of aggregate temperatures in summer and early autumn. But, when comparing temperatures with Oechsle ratings, we must remember that quantity interferes somewhat with quality, i.e. the higher the yields per acre the lower the sugar content will be under the same climatic conditions. For this reason the use of standardized Oechsle ratings (°Oe at kg/m²) is recommended. Basler has found very high correlations in different wine growing regions of eastern Switzerland (R^2 of .92 to .95) between the temperatures at noon above 12° to 15°C aggregated from June to the date of the harvest. The correlation with August, however, was found to be significantly lower than for the other months.[30]

In the past the qualities were described in such terms as mediocre, excellent, and bad. Although we cannot be sure that the absolute standards of taste have not changed over the centuries, comparisons have shown that these subjective estimates of quality do agree rather well with the Oechsle ratings. Unfortunately most wine chronicles are non-contemporary for most of the period covered; the famous work of Müller is, for instance, not entirely based on verified sources and should be used with great care. In addition, Weise points out the large variance from one vineyard to another and recommends against analyses of isolated series.[31]

29 Nilolaus Weger, "Weinernten und Sonnenflecken," *Berichte des Deutschen Wetterdienstes in der US-Zone,* 38 (1952), 229–237; Alessandro Rima, "Considerazioni su una serie agraria bisecolare; la produzione di vino nel Rheingau (1719–1950)," *Geofisica e Meteorologia,* XII (1963), 25–31; Peter Wright, "Wine Harvests in Luxembourg and the Biennial Oscillations in European Summers," *Weather,* XXIII (1968), 300–304; Karl Müller, *Geschichte des Badischen Weinbaus* (Lahr, 1953).

30 At the beginning of the nineteenth century Ferdinand Oechsle from Pforzheim invented an instrument to weigh the sugar content of wines. The degrees Oechsle (°Oe) indicate the specific weight of the grape juice (SWJ): SWJ = °Oe + 1000. The sugar content in grams equals about twice the Oechsle rating. A wine of 80° therefore has a specific weight of 1080 g and a sugar content of 160 g. Basler, "Beeinflussung"; Becker, "Oekologische Kriterien," 80; Hofäcker, *Einfluss von Umweltfaktoren,* 36–38; Müller, *Weinbaus,* 241; N. E. Davis, "An Optimum Summer Weather Index," *Weather,* XXIII (1968), 305–318; Rudolf Weise, "Ueber die Rebe als Klima Kriterium," *Berichte des Deutschen Wetterdienstes in der US-Zone,* 12 (1950), 121–123.

31 Wright, "Luxembourg," 302; Edmond Guyot and Charles Godet, "Le climat et la vigne," *Bulletin de la Société des Sciences Naturelles de Neuchâtel,* LX (1935), 218. Weise, "Rebe als Klima Kriterium," 123.

Wine quantity. Yields per acre are not only affected by climate but, in the long term, also by a change in varieties, manuring, cutting, and the technique of cultivation. Climatic factors, together with diseases, may only help to explain the short-term fluctuations from year to year. Detailed analyses of yield series have been made in order to determine the weight of the different components which may affect the size of the harvests.[32]

Based on a period of sixty-two years (1871–1932) for which the mean yield per acre in the Canton of Vaud (western Switzerland) was available, Guyot and Godet found that temperatures in July and August significantly affected yields. If these months were hot and sunny and at the same time the plants could find enough water in the soil, a plentiful yield could have been expected. If cold and wet spells occurred in June and early July, they affected the flowering. Similar conditions in July and August affected the growth of the grapes as well over very large areas. The obvious damage caused by late frost and hail, however, varied considerably from one vineyard to another and was less significant, as can be seen from an analysis of the yield series of several wine growing regions. In addition, widespread and heavy damage by frost, such as occurred in 1709 and 1740, was always reported by a great number of weather chroniclers and vine growers and can therefore be considered in the interpretation. Also the burgeons that were killed by late frost were often replaced by supplementary burgeons. We may therefore conclude that a careful analysis of several series of wine yields from rather distant areas may well reveal fluctuations of temperatures in the summer months.[33]

32 Hans Schwarzenbach, *Die Produktivitätsentwicklung im schweizerischen Weinbau* (Bern, 1963), 11; Bernard Primault, "Le climat et la viticulture," *International Journal of Biometeorology*, XIII (1969), 7–24.

33 Guyot and Godet, "Le climat et la vigne," 209–223; Guyot, "Calcul de coefficients de corrélation entre le rendement du vignoble neuchâtelois, la température et la durée d'insolation," *Bulletin de la Société des Sciences Naturelles de Neuchâtel*, LXV (1940), 5–15; Koblet, "Fruchtansatz bei Reben in Abhängigkeit von Triebbehandlung und Klimafaktoren," *Weinwissenschaft*, XXI (1966), 297–323; Primault, "Climat et viticulture," 17. Müller, *Weinbaus*, 244, points to the fact the several diseases, in particular infection with the peronospora fungus, was only observed in wet summers. Weger, "Weinernten und Sonnenflecken," 234. The burgeons survive temperatures of −3.5°C unless they are dry; wet burgeons are killed at −1.5°C, especially through hoarfrost. The plant also stands

In contrast to grain harvests, which are increasingly well documented through tithes, little is known historically about wine production. This is probably because the annual fluctuations in grain harvest yields are more conclusive for a historian, who is concerned with demography, agrarian legislation, and the social impact of subsistence crises. Although wine was also an important staple food, substitute beverages were readily available.[34]

In what follows I outline and discuss the climatological significance of the evidence drawn from several Swiss archives.[35]

In most counties of the two protestant states of Bern and Zürich, which were either situated in vine-growing areas or had inherited possessions from ancient religious orders, a part of the governors' revenues was paid in wine. In some cases the nature of the payments was not specified in the sources; more frequently, we know that they came from vineyards which were cultivated by individual vine-growers in return for a third or half of the yield. The governor in his turn took the other half or two thirds of the harvest. This suggests that those revenues directly reflected variations in yields. A sample of several local series was taken from each major vine-growing region on the Swiss plateau. Subsequently, four regional series and a main series were aggregated from these data.[36]

Occasionally, the size of the individual vineyards, from which the payment came, was specified in a source, which allows computation of yields per acre. The longest series of that kind (1538–1838) could be drawn from the account books of the ancient foundation of Fraumünster in Zürich. From the data for which the size of the vineyard was specified in the source, a series of mean yields per acre was computed.

---

extreme winter temperatures up to −20°C when the cold spell sets in early enough. Primault, "Climat et viticulture," 9; Alfred Schellenberg, *Weinbau* (Frauenfeld, 1966), 94ff.

34 Joseph Goy and Le Roy Ladurie (eds.), *Les fluctuations du produit de la dîme. Communications et travaux* (Paris, 1972). A considerable number of new tithe series from many European countries were presented at the conference on "Prestations paysannes, dîmes, rente foncière et mouvement de la production agricole" (Paris, 1977).

35 Pfister, "Die Fluktuationen der Weinproduktion im Schweizer Mittelland vom 16. bis ins frühe 19. Jahrhundert," *Schweizer Zeitschrift für Geschichte*, XL (1980), forthcoming.

36 *Ibid.* The main series was composed from the residuals of the local series, which could all fairly be described through linear trends. It was assumed that those trends included mainly changes in the surface cultivated and long term changes in yields per acre.

A comparison of the main series of yields, for which the size of the vineyards was not known, with the series of mean yields per acre produced a highly significant correlation coefficient of 0.85. Thus, we may conclude that the former series for the most part reflects fluctuations of yields per acre. Also, it turned out that the residuals of the regional series were significantly correlated among each other as well as with the individual series of yields per acre. This is consistent with the observations of Müller and Weise, who noted that in Germany years of plenty and years of dearth mainly agreed, even between rather distant vine-growing regions.[37]

The Basel series was again used for the comparison of the wine yields with the temperatures of the individual months. A stepwise regression analysis yielded that July had the greatest weight, with $r = .43$, which is consistent with the result obtained from recent data, where r was .46. Taken together, the temperature of the three summer months could explain 46 percent of the variance in the yield curve. The hypothesis that wine years from the historical past are valid indicators of summer temperatures was also confirmed through the significant correlations with the tithe auction dates and the wine harvest dates.[38]

The trends of the curve in Figure 3 reflect the major phases of the little ice age: the long term drop over the six decades from 1530 to 1600 took, after 1560, the form of clusters of bad years, which become increasingly long, frequent, and pronounced. Among the thirteen years from 1585 to 1597 all except one (1593) provided yields below the long term trend; in the worst cases (1588 and 1589) the grapes collected in some vineyards hardly filled a hat. This patch of cold wet summers also saw quick and far-reaching advances of the alpine glaciers during the 1590s. Another cluster of bad wine years (1618–1629) preceded a wave of glacial activity after 1630.

From 1630 the curve climbs steadily up to the turning point in 1687, which introduces the barren 1690s. During the first two

37 Coefficient of correlation $r = .85$ for 266 paired values. Significance 0.00001. A correlation matrix is provided by Pfister, "Weinproduktion." Müller, *Weinbaus*, 240; Weise, "Rebe als Klima Kriterium," 122.

38 Guyot and Godet, "Le climat et la vigne." Vine yields correlated with tithe auction dates: coefficient of correlation $r = .49$ for 187 paired values. Significance .0001. Vine yields correlated with Swiss wine harvest dates: coefficient of correlation $r = .36$ for 187 paired values. Signifiance .001.

Fig. 3 Fluctuations in Wine Yields at the Swiss Plateau, 1530–1825[a])

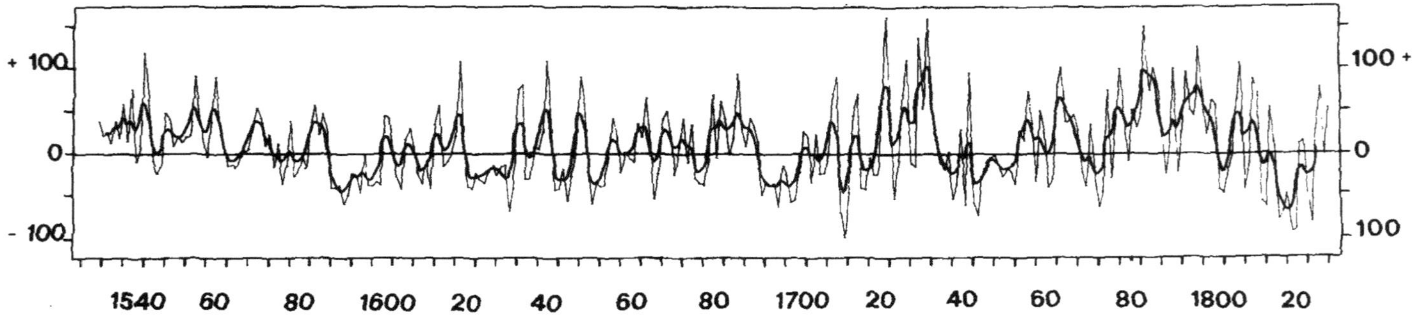

SOURCE: Pfister, "Fluktuationen."

a Average of 14 local series from the major vine-growing regions, represented as deviations from the mean (%). The bold line marks a five-year running mean.

decades of the eighteenth century the trend was uneven. Suddenly, from 1719 to 1729 the vine became exuberant. From Lake Geneva to the Rhineland wine casks overflowed; the glaciers, after a secondary peak around 1720, retreated. In the following five decades a change to moderate slumps and peaks becomes observable and can clearly be connected to minor episodes of glacial history. The abundance of wine around 1780, comparable to that of the 1720s, was also witnessed in the vineyards of France and Germany. From 1794 the curve descends to its lowest point in 1816, the year without a summer. Again, in the following years, quick and far-reaching glacial advances occurred.[39]

Although the fluctuations of vine yields and wine harvest dates were roughly parallel, some inconsistencies should not be overlooked. Sometimes, notably in the second half of the 1580s and of the 1620s, the slump in vine yields was more marked than the delay of wine harvests. Given the presumption that wine harvest dates were strongly responsive to temperatures in spring and early summer, whereas vine yields were chiefly indicators of temperatures in midsummer, in both periods midsummers may have been notably colder than springs and early summers. For 1588, 1589, 1622, and 1625, where the differences were most striking, this pattern can be supported by descriptive evidence.[40]

The wine boom of the 1720s and the 1780s, however, was not accompanied by a proportionate advance of wine harvest dates. Three tropical summers occurred in both decades (1723, 1727, and 1729, 1781, 1783, and 1788). They were all very warm with frequent thunderstorms, which stimulated the yields of wine far beyond normal. The wine boom of those years may also be explained by the fact that the previous summers were warm. According to experts in vine-growing, warmth increases the number of flowers in a following spring.[41]

Thus wine yields, like tithe auction dates, not only support the evidence of the wine harvest dates; if they are properly mar-

39 Rima, "Produzione di vino nel Rheingau," 26, Tisowksy, "Häcker und Bauern in den Weinbaugemeinden am Schwanberg," *Frankfurter Geographische Hefte, XXXI* (1957), 50; Heinz J Zumbühl, *Die Schwankungen der Grindelwald-Gletscher inden historischen Bild- und Schriftquellen des 12. -19. Jahrhunderts* (Zürich, 1980); Le Roy Ladurie, *History of Climate,* 207.

40 See CLIMHIST data bank.

41 Müller, *Weinbaus,* 240

shalled and studied in their ecological context, they may give us more detailed knowledge of the character of past summers. Such knowledge may then help to explain some of the inconsistencies which still exist between the evidence from glacial and from proxy data.

In addition, the analysis of wine yield series may also be conclusive for the economic history of the vine-growing regions of Central Europe. The evidence suggests that wine-growers lived through the parallel coincidence of good and poor yields, a common economic experience, which was closely tied to climatic history.

THE COMPILATION OF INDICES Climatic history may be analyzed in two ways. The first approach relates field data to a complex of climatic variables, and traces their fluctuations back in time. Although the time resolution of some data may be high, it is always restricted to a specific season or interval. In addition, most field data and historical proxy data respond to a complex of meteorological variables, chiefly rainfall and temperatures, which are difficult to disentangle. Also, the response is in most cases a function of the weather patterns during several months.

The second approach, which should rather be called weather history, analyzes different kinds of descriptive evidence, very accurately dated and related to specific meteorological events, which can only occasionally be used to estimate temperature or rainfall. One form of analysis of this fragmentary type of information is to transform the material into a numerical index prior to interpreting it in terms of standard meteorological variables. Indices of wetness and winter severity were first derived by Brooks and Easton in the 1920s. After 1960 this work was refined by Lamb, who computed decadal indices via compilations from chronicles, annals, diaries, and similar sources.[42]

The present approach attempts to bridge the gap between climatic history and weather history by cross-dating different kinds of field data (such as tree ring densities and glacial advances) and documentary proxy data (such as phenological observations,

42 C. E. P. Brooks, *Climate through the Ages* (London, 1926); C. Easton, *Les hivers dans l'Europe occidentale* (Leyden, 1928); Lamb, *Climate*, 34–5, 440; Pierre Alexandre, *Le Climat au Moyen Age en Belgique et dans les Regions Voisines (Rhenanie, Nord de la France)* (Liège, 1976).

paraphenological data, and wine yields) with the large number of descriptive records in the CLIMHIST data bank, in order to obtain rough estimates of temperature and wetness for individual months. Two types of indices are derived, a weighted and an unweighted one. For the construction of a *weighted thermal index* and a *weighted wetness index,* which are described in more detail elsewhere, weight factors ranging from +3 to −3 have been applied. In order to derive the thermal index proxy data were used to assess the magnitude of a temperature deviation from the long term mean, whereas the exact timing of the corresponding warm or cold weather spell was obtained from descriptive evidence. The wetness index is based upon the number of rainy days counted in weather diaries, and descriptions of floods and droughts given in chronicles and rainfall measurements (from 1708).[43]

The *unweighted decennial thermal index* in Figure 4 gives the excess number of unmistakably mild months (M) per decade contrasted with months of unmistakably cold character (C), i.e. M–C, for the individual months and the entire seasons of the spring–summer period. Correspondingly the *unweighted wetness index* is W–D, where W is the number of months with evidence of frequent rains and D is the number of months with evidence of drought per decade. In both cases unremarkable months and those without observations score zero. The "difference" index was used despite its known sensitivity to missing data, because the number of spring and summer months without any observation was very small (about 20 out of 300 for each month).[44]

The second step in indexing is to convert the crude index to a meteorological parameter. For the thermal index it was assumed that the unmistakable cases were at least 1°C warmer or colder than the 1901–1960 average. During the last six decades covered with thermometrical measurement, the months were classified according to the same criterion, and the index in Figure 4 was also compared with the measured decennial averages. The figure illustrates that both curves are parallel for all months and that two points of the index represent roughly 1°C.

43 Pfister, "Swiss Historical Weather Documentation"; *idem,* "Die älteste Niederschlagsreihe Mitteleuropas; Zürich 1708–1754," *Meteorologische Rundschau,* XXI (1978), 56–62.
44 Ingram and Underhill, "Use of Documentary Sources," 81.

Fig. 4 Decennial Unweighted Thermal and Wetness Index for Spring and Summer (Monthly and Seasonal Values)

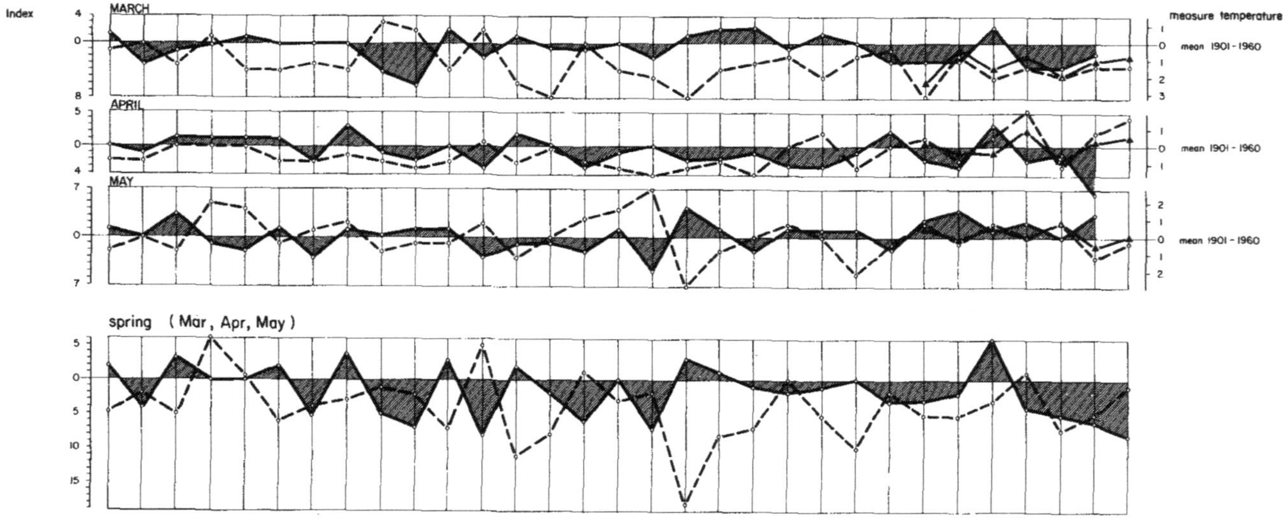

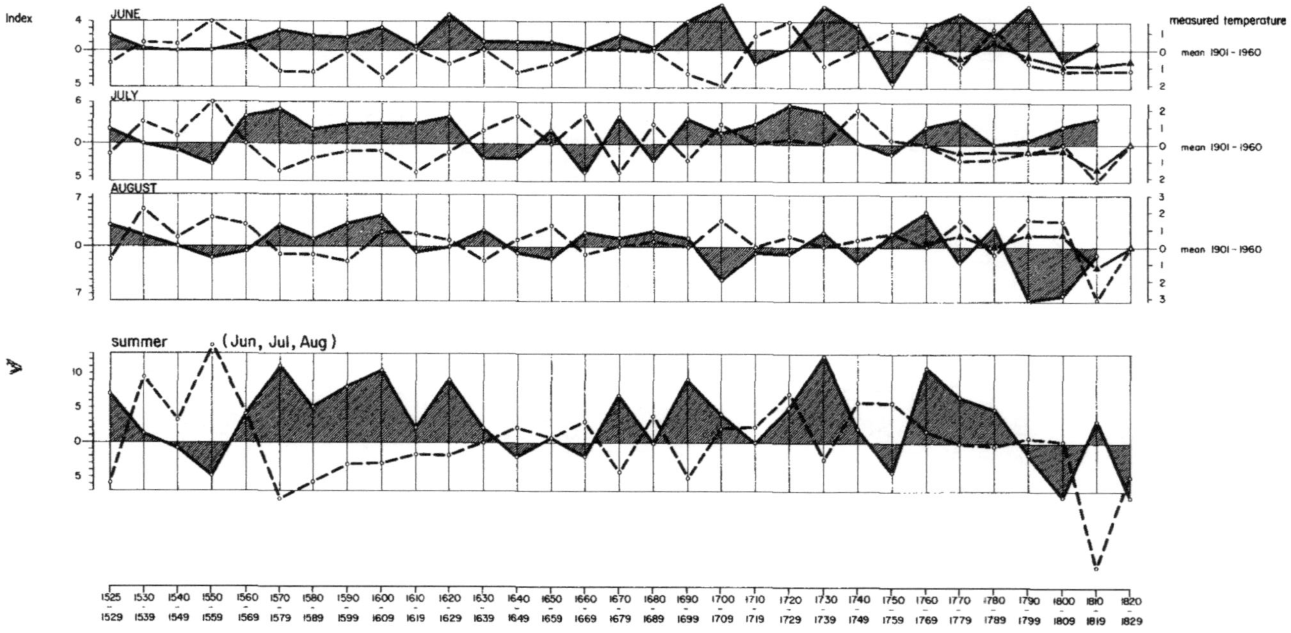

thermal index — wetness index — measured temperature

The excess or deficit of wet months is set off with the hatched area

For the wetness index the measured precipitation or the number of rainy days of each month was compared with a histogram of the 1901–1960 precipitation statistics at the same place. Months below the lower quartile range (at least as dry as the fifteen driest months) were classified dry; those above the upper quartile range (at least as wet as the fifteen wettest months) scored wet.[45]

THERMAL AND WETNESS INDICES FOR SPRING AND SUMMER, 1525–1825 Figure 4 displays decennial means of unweighted thermal and wetness indices for each individual spring and summer month and for both seasons. The wetness curve is set off with the tinted area; the broken curve represents the thermal patterns. Although the thermal curve is mainly based upon observations on the plateau, it also represents conditions at higher altitudes, because temperature patterns in spring and summer (but not in autumn and winter) are highly correlated between lowland and upland stations.[46]

The character of the spring months. Of all the months represented March reflects the climatic change of the little ice age most persistently and distinctly. Until 1560 no clear pattern emerges. Then the cooling starts. March remains cold, in many cases a real winter month, for the entire period displayed in the figure (with the noticeable exception of the two warm and dry decades at the beginning of the seventeenth century).

The coldest decade documented with measured evidence (1760–1769) was 2° below the 1901–1960 mean; a similar deviation has been estimated for the 1690s based upon careful observations of snowcover in a weather diary from Zurich. During the 1640s, likewise, March was probably just as cold. This persistent cold, together with the prevailing notion of drought (fifteen decades with an excess of dry months as against eight with an excess of wet months), suggests that this month was frequently dominated by northerly winds and blocking anticyclones. A significant warming in March did not occur before 1900.[47]

45 If all observations are rank ordered from highest to lowest, the 25th percentile corresponds to the lower quartile and the 75th percentile to the upper quartile.

46 Paul Messerli, *Beitrag zur statistischen Analyse klimatologischer Zeitreihen* (Bern, 1980).

47 Bider et al., "Basler Temperaturreihe," Table 1; Pfister, "Klima des Raumes Zürich," passim. Kington, "Historical Daily Synoptic Weather Maps from the 1780s," *Journal of Meteorology*, III (1978), 65–71, presents a synoptic analysis of this extremely cold March in 1785.

April shows less variation than March. From 1525 to 1720 the thermal index fluctuates slightly below the zero line, which suggests that this period, taken as a whole, was probably somewhat below the 1901–1960 average. After 1740 a warming of April is observable, which is maintained until the end of the little ice age. Although the wet months are more frequent during the sixteenth century, April was rather dry during most of the seventeenth and eighteenth centuries (with the exception of the 1640s, the 1750s, and the 1780s).[48]

May does not reflect a typical little ice age pattern. If we take the period from 1525 to 1825 as a whole, neither temperature nor precipitation seems to have significantly differed from the mean of the present century. After the cold wet decade of the 1540s, there was a net excess of warm and dry months up to the turn of the century, notably during the 1550s and the 1560s. During the first half of the seventeenth century, warm and cold months roughly balance. Then, from the 1660s the curve rises to its highest peak of +7 in the 1680s. At the same time the wetness index goes down to a minimum of −5. In sharp contrast, the 1690s show a drastic slump in the thermal index (−14 points) which is accompanied by a sudden increase in wetness (+9 points). In central England the measured difference between these two decades was 1.8°; in Switzerland it may have been considerably greater. Another marked excess of cold months stands out in the 1740s.

The polarity between the warm period from 1580 to 1690 (with an excess of 12 index points) and the succeeding cold period from 1690 to 1800 (with a deficit of 13 index points) could account to some extent for the delay of the wine harvests in the eighteenth century compared with those in the seventeenth century, given the sensitivity of the growth pattern of grapevines to temperatures in May (see Table 2).[49]

If we look at the composite picture of the three spring months, the dominant impression is one of coldness and drought. Only two decades (the 1550s and the 1630s) had an excess of more than one warm month, whereas twelve decades were −5 or be-

48 Bider et al., "Basler Temperaturreihe," Table 1.

49 Gordon Manley, "Central England Temperatures: Monthly Means 1659 to 1973," *Quarterly Journal of the Royal Meteorological Society*, C (1974), 389–405. Le Roy Ladurie and Baulant, "Grape Harvests"; Anne-Marie Piuz, "Climat, récoltes et vie des hommes à Genève, XVIe -XVIIIe siècle," *Annales*, XXIX (1974), 599–618.

low, the coldest being the 1690s (−18), the 1640s (−11), and the 1740s (−10). These tendencies are clearly reflected in the wine harvest dates. An analysis of the Basel temperature series by Messerli has revealed that for the period 1860–1965 springs were .3° warmer than in the preceding period 1755–1860, which is statistically significant.[50]

The character of the summer months. June clearly reflects the climatic change of the sixteenth century. Around 1560 this month, which had been predominantly warm during the preceding two decades, became wet and cool. Most remarkably, this tendency was not interrupted for the next 140 years (except during the 1660s, when both indices were at zero). By contrast, the excess of cold months was always moderate; the maximum (−5) was observed in the 1700s together with the greatest excess in wetness (+7). During the eighteenth and early nineteenth centuries there was an alternation of periods which were warm and dry (1710s, 1720s, and 1750s) with periods which were rather cool and wet (1730s, 1740s, 1760–1800, and 1810s). As a whole, the predominance of wetness prevails.

July presents a very similar picture until 1630: hot and dry midsummers from 1540 to 1560 were followed by an uninterrupted series of seven damp and cool decades. From 1630 to 1670, however, Julys were clearly warmer and drier than Junes. Also during the 1690s the excess of cold and wet months was less pronounced. Again, from 1700 to 1760 the thermal index was above or at the line, whereas wetness was often very pronounced (+6 in the 1730s). After 1760 a second long succession of cool and rainy decades set in which culminated in the 1810s, when the July curve descended to its lowest point (−5). This corresponds to a deviation of minus 1.2° from the 1901–1960 mean. The slumps of 1570–1579, 1610–1619, and 1670–1679 (−4 each) might have been of an order of magnitude of −1°C. On the other hand, the peak of the 1550s (+6) may have been equal to the warmest decennial mean for July (1943–1952), which, in Central Europe, was about 1.2° above average.[51]

August, taken as a whole, was probably no colder and wetter

50 Paul Messerli, *Beitrag,* 24.
51 Bider et al., "Basler Temperaturreihe," Table 1.

than in the present century. In contrast to the preceding months, the warm and dry phase of the mid-sixteenth century was extended to the 1560s, and from 1600 to 1630 late summers were rather warm. Thus the cool and wet pattern of Junes and Julys at the end of the sixteenth century did not include August, except during the three decades from 1570 to 1600. Similarly, the coldest phase of the little ice age, the 1690s, is not reflected in this curve. Again, from 1790 to 1810, the cold and wetness of June and July were counterbalanced by the warmth of August.

A composite curve combines the composite thermal and wetness patterns of the three summer months. The significance of the climatic change which occurred during the sixteenth century is revealed by a comparison of the number of "warm," "cold," "dry," and "wet" summer months from 1525 to 1600:[52]

1525–1569		1570–1600	
warm	*cold*	*warm*	*cold*
48	21	26	44
dry	*wet*	*dry*	*wet*
24	30	11	35

From 1560 to 1630, taken as a whole, summers were persistently cool and wet. Although cooling was most pronounced at the beginning (−8) and diminished afterwards, the excess of wet months was mostly above +5, except from 1610 to 1619. If we allow for a time lag, this pattern is clearly connected to the thrust of the alpine glaciers from the mid-1580s to the turn of the century. From 1630 to 1670 both indices are near zero, which suggests that summers were roughly equal to those of the present century. During the little interglacial, as we might call that episode, most glaciers began to retreat. From 1670 to 1740, and again from 1760 to 1790, there was a clear excess of wet summer months, although the summers were not so frequently cold as during the first phase of the little ice age (1560–1630). This may explain why the dimension of the glacial advances during that time remained limited.

The long-lasting advanced position in the nineteenth century was chiefly triggered by the short-term fluctuation of the 1810s,

52 Pfister, "Swiss Historical Weather Documentation."

the most drastic slump in temperature (−17) contained in the index. Many glaciers, however, were already somewhat advanced when the fluctuation began.[53]

It has been demonstrated that crossdating instrumental data, field data, documentary proxy data, and a body of descriptive data yields a refined picture of the thermal and wetness patterns in spring and summer. What can we conclude from the result?

When we reduce our focus to the level of individual seasons and months only two features persist through the entire period of the little ice age: the cold in March and the cool and wet character of June. If we consider the whole pattern, the little ice age in Central Europe was a period of rather heterogeneous climate. As far as the spring-summer period is concerned, like the big ice age, it crumbles away into a variety of sub-periods, into minor fluctuations, which might be called little interglacial or little little ice age, if the signs of the indices agree in several seasons for more than a decade. On a shorter time scale three periods stand out, in which springs and summers were simultaneously cold: 1570–1600, the 1690s, and the 1810s. This conclusion holds not only for Switzerland, but also for large parts of the European continent and probably for the northern hemisphere as a whole. As we may conclude from a variety of studies, those phases of climatic hardship had, particularly in the marginal regions, severe economic and demographic impacts upon many societies of preindustrial Europe.[54]

53 Le Roy Ladurie, *History of Climate,* 207; Zumbühl, *Grindelwald-Gletscher*; Bruno Messerli et al., "Schwankungen," passim.

54 Tom M. L. Wigley et al., "Geographical Patterns of Climatic Change: 1000 B.C.–1700 A.D.," *Quaternary Research,* forthcoming; Piuz, "Climat, récoltes"; Gustav Utterström, "Climatic Fluctuations and Population Problems in Early Modern History," *Scandinavian Economic History Review,* I (1955), 3–47; Francois Lebrun, *Men and Death in Anjou in the Seventeenth and Eighteenth Centuries* (Paris, 1971); K. Walton, "Climate and Famines in Northeast Scotland," *Scottish Geographical Magazine,* LXVIII (1952), 13–21; John D. Post, *The Last Great Subsistence Crisis in the Western World* (Baltimore, 1977).

Jerome Namias

Severe Drought and Recent History

Man's existence throughout history has been plagued by drought. Areas with either moist or dry climates have suffered from drought in normal times or in periods of climatic extremes. Droughts are not isolated events but are part of the large-scale patterns of atmospheric circulations that determine climate in their normal configurations but can lead to drought and flood in their extreme deviations from normal. The configurations of atmospheric flow tend toward certain fixed sizes or wave lengths and thus we have "teleconnections"—a meander of large-scale flow in the atmosphere over one area of the earth's surface tends to be associated with repercussions in other areas. For example, if a deep trough with stormy weather lies over the North Pacific in proper position to be associated with a ridge over land, then the area dominated by the high-pressure ridge will be dry. If the condition is prolonged or oft-repeated and if it catches the natural water supply at a low ebb then drought ensues. These teleconnected events often have positive feedbacks and are synergistic in character.

Over the years these abnormal patterns repeat themselves—not periodically so that we can predict them, but at seemingly random intervals. Some places are more natural targets for drought, although changes in climate can shift the positions of these target areas, but drought remains an enduring problem.

The definition of drought is partly contingent upon its impact on society and on the economy. For the purposes of this article we shall consider drought as an extended period of deficient precipitation relative to normal. According to this definition drought can occur almost anywhere in the world, because the natural variability of precipitation is a reliable statistical characteristic of climate.

Jerome Namias is Research Meteorologist and Head of the Climate Research Group at Scripps Institution of Oceanography.

Charles K. Stidd and Dan Cayan assisted the author in editorial and other matters associated with the preparation of this article, and their help is gratefully acknowledged. Carolyn Heintskill typed the manuscript. Research, of which this paper is a partial result, was supported by the National Science Foundation, Office for the International Decade of Ocean Exploration under Contract OCE78-25132 and the Office for Climate Dynamics under Contract ATM78-24003. The author also acknowledges the partial support of Project Hydrospect of the State of California, Department of Water Resources.

In this article we are concerned primarily with drought in temperate latitudes. The following material focuses on the statistical and physical aspects of drought on time scales of a month to several years. There are many unsolved "mysteries" of drought; although some physical understanding has been achieved for droughts that last from a month to a season, spells of years characterized by drought are poorly understood, and thus remain on the agenda of research climatologists.

Drought is the statistical aggregate of persistent and persistently recurrent meteorological events. Thus, if a high-pressure area and associated dryness enters a given region for a few days, a similar weather situation is apt to recur several days later, and recur again and again. The net effect is to produce, in wind and weather patterns, deviations from the mean that are highly anomalous. The central question, therefore, is why these drought producing patterns have an affinity for certain areas at certain times. We see below that the drought problem is not local in character but, rather, is almost global because of interconnections among parts of the general circulation of the atmosphere.

The literature on drought is voluminous. The Russians have made a number of studies of droughts and *sukhoveis* (foehn winds) in their European territory in order to understand the mechanics of such phenomena and their effect upon agriculture. There is a drought bibliography containing hundreds of references to articles on drought and regional studies have appeared by the thousands.[1]

During the 1970s the subject of drought has received attention from the news media and the scientific community because of such severe events as the Russian drought of 1972, the two-year, back-to-back drought of 1976 and 1977 on the west coast of the United States, the European drought of 1976, and the Sahel drought of 1972–73.[2]

1 B. L. Dzerdzeevskii (ed.), "Sukhoveis and Drought Control," *Izdatel'stvo Akademii Nauk SSSR* (Moscow, 1957). [Trans. from Russian by Israel Program for Science Translations (Jerusalem 1963)], 366; Wayne C. Palmer and Lyle M. Denny, "Drought Bibliography," *National Oceanic Atmospheric Administration Technical Memorandum, Environmental Data Service,* XX (1971), 236. For examples of regional studies, see Donald G. Friedman, "The Prediction of Long-continuing Drought in South and Southwest Texas," *Travelers Weather Research Center, Occasional Papers in Meteorology* (Hartford, 1957), 182; California Department of Water Resources, *The 1976–77 California Drought: A Review* (Sacramento, 1978), 228.

2 A. L. Katz (trans. Lydia A. Hutchinson), *The Unusual Summer of 1972* (Leningrad, 1973), 58; Namias, "Multiple Causes of the North American Abnormal Winter 1976–77," *Monthly Weather Review*, CVI (1978), 279–295; *idem*, "Recent Drought in California and

PHYSICAL FACTORS ASSOCIATED WITH MID-LATITUDE DROUGHT

The most characteristic signature of regional drought over most areas of temperate latitude is the presence of warm, dry air in the middle troposphere. This warmth aloft can be associated with slow sinking motions (subsidence) of the order of several hundred meters per day. Effectively, the sinking motions and associated adiabatic heating and low relative humidity inhibit precipitation, because it is ascending air motion with attendant adiabatic cooling and condensation that is responsible for most precipitation. Besides, during warm seasons dry and warm air aloft discourages the growth of cumulus clouds, both because of the static stability of the air and because of the entrainment of moist cloudy air with the dry ambient air.[3]

The phenomenon of subsidence was first described and quantified by Margules, an eminent meteorologist. He showed not only that sinking motions led to warming of the air adiabatically and to stabilized lapse rates (changes of temperature with altitude), but also that during subsidence there frequently were horizontally diverging air masses that led to even more stable lapse rates. Figure 1 shows an example of the effect of subsidence on the vertical temperature distribution, in this case in the core of the severe drought in Great Britain during the summer of 1976. The relative warmth aloft can be accounted for by subsidence of roughly several hundred meters per day. The deviation in rainfall is given in the lower left hand corner of this chart.[4]

What features of the general atmospheric circulation are responsible for subsidence and why is there persistent recurrence of these drought-producing features? Subsidence, particularly during warm seasons, is associated with high pressures mainly at upper levels but at times at the surface. Because of surface friction, the

Western Europe," *Reviews of Geophysics and Space Physics,* XVI (1978), 435–458; M. K. Miles, "Atmospheric Circulation during the Severe Drought of 1975–76," *Meteorological Magazine,* CVI (1977), 154–164; R. A. S. Ratcliffe, "A Synoptic Climatologist's Viewpoint of the 1975–76 Drought," in *ibid.,* 145–154; Helmut E. Landsberg, "Sahel Drought; Change of Climate or Part of Climate?" *Archiv für Meteorologie Geophysik und Bioklimatolgie,* XXIII (1975), 193–200.

3 Adiabatic means without the addition or removal of heat. The air as it descends is heated by compression because of the increase in pressure. Warm air is less dense than cold air and if the air at high intervals is warm, the condition is stable, i.e., there is little tendency for overturning or vertical mixing. If relatively warm air is near the surface and relatively cold air aloft, an unstable condition exists, often leading to ascending motions and heavy showery rainfall.

4 M. Margules, "Zur Sturmtheorie," *Meteorologische Zeitschrift,* XXIII (1906), 381–497.

Fig. 1 Upper Air Temperature in the Core (Crawley) of the British Drought in June, 1976, Relative to the More Normal June, 1974 Temperature.[a]

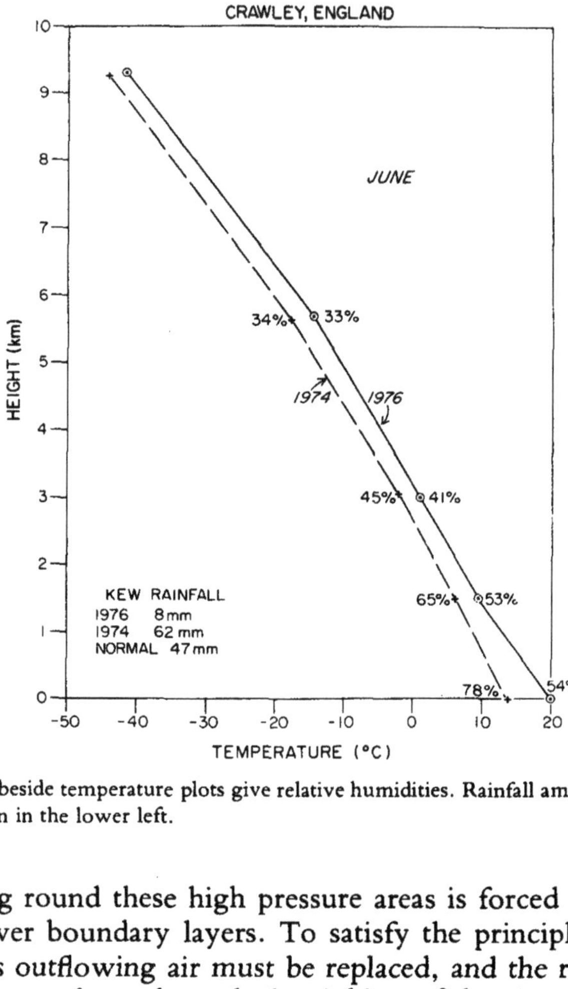

a Numbers beside temperature plots give relative humidities. Rainfall amounts at nearby Kew are given in the lower left.

air flowing round these high pressure areas is forced to leak out of the lower boundary layers. To satisfy the principle of continuity, this outflowing air must be replaced, and the replacement usually comes about through the sinking of the air masses aloft.

But this frictional divergence is not the only cause of subsidence. The dynamics of atmospheric long waves in the westerlies (Rossby waves) demands areas of convergence and divergence of air aloft. The converging air results in increased pressure aloft that is frequently compensated by divergence and lowering pressure below. The piling up of air aloft is frequently found in the deep warm anticyclones associated with drought. Frequently, par-

Fig. 2 Sea-level Pressure and Departures (mb) from the Long-term Mean for Winter 1975–1976.

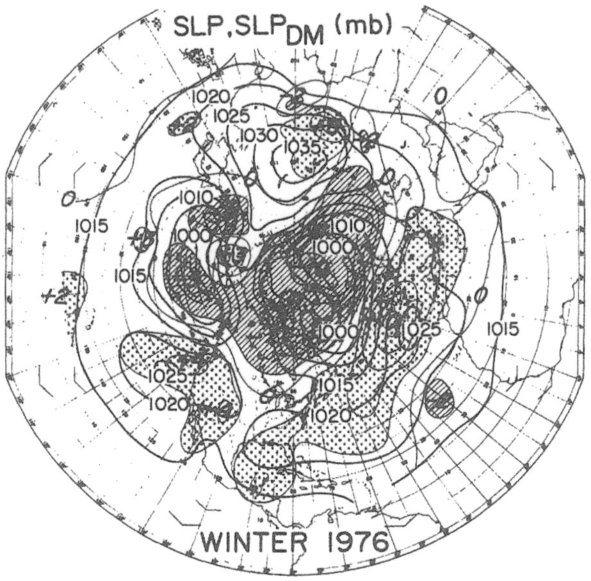

ticularly during cold seasons of the year, sinking motions are induced in the high-level northwest currents of air behind long-wave troughs. Thus, in these regions, if the flux is sustained or recurrent, substantial deficiencies in precipitation, and hence drought, can occur.[5]

In addition to the dynamics and thermodynamics associated with subsidence, one must consider the source and trajectory of moisture. Tongues of moist air emanating, let us say from the Gulf of Mexico, given proper systems involving ascending motion, can produce heavy rains, whereas descending dry tongues, flung southward from continental sources, can lead to deficient precipitation.

An example of what we have described can be seen in Figure 2, which shows the pressure distribution and its standardized anomaly associated with the California west coast drought of winter 1975–76 and the beginning of the European drought. Anticyclonic conditions and positive anomalies of pressure dominate

5 Jacob A. Bjerknes, "Theorie der aussertropischen Zyklonenbildung," in *ibid.*, LIV (1937), 462–466.

Fig. 3 Monthly Average 700 mb height, Temperature Departure from Normal, and Isentropic Chart for August, 1936.

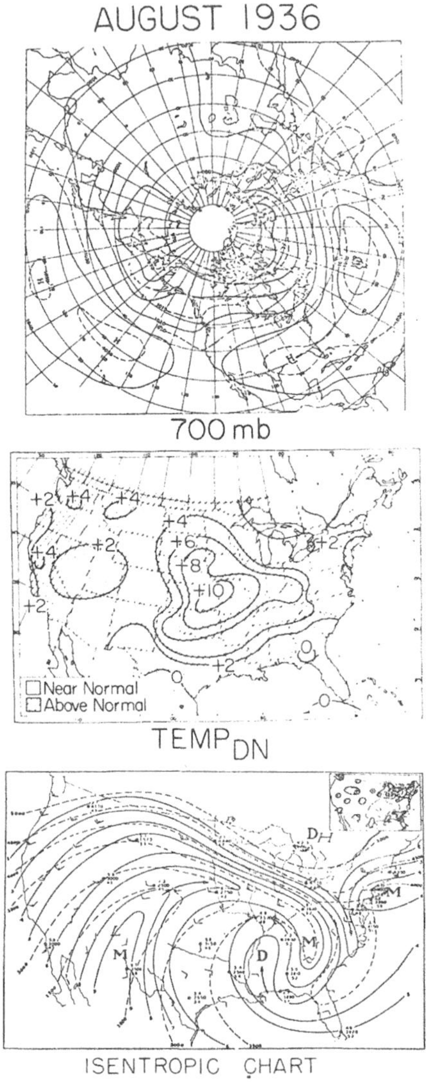

the drought areas. Figure 3, one of the months of the especially severe Dust Bowl drought of 1936, shows the upper-level anticyclone and the associated extreme warmth over the central plains of the United States. The warmth was in part associated with the lack of cloud and precipitation during this period, and the consequent increase in sunshine. Figure 3 also shows the moist (M) tongues and (D) dry tongues associated with the 1936 case. Over the central and southern plains of the United States, dry air originally from Canada spirals into the drought-affected area, sinking as it progresses. In the upper right of the isentropic chart in Figure 3 are the deviations from normal of precipitation. Strong upper level anticyclones with concomitant warm, dry air are common to most drought situations, as for instance in the recent Russian drought in the summer of 1972, the California drought in the winters of 1976 and 1977, and the British drought during the summer of 1976.[6]

There are oceanic high-pressure cells accompanying the high-pressure areas over the drought-affected areas. For example, in the 1936 drought (Fig. 3), strong high-pressure areas existed over the Atlantic and Pacific Oceans as well as over the continental United States. Each drought-producing cell appears to require for its sustenance companion cells remote by thousands of kilometers. These cells are the great centers of action first described by Teisserenc de Bort in 1881. Their explanation as reflections of the mid-tropospheric prevailing wind patterns was first given by Rossby and his collaborators in 1939. These high-pressure areas (upper-level ridges) together with the low-pressure troughs comprise the upper-level westerlies. The positions of these long waves determine to a large extent climate and variations in climate. Drought is the statistical manifestation of some aberrant form of the long waves, particularly of changes in their position and amplitude relative to normal.[7]

Because the high-pressure centers-of-action over the hemisphere are interrelated or teleconnected, there may be three to five of these anomalous high-pressure cells present in any one season. This is especially true during the warm season when the prevailing westerlies are farthest north. Thus, if we were to average the zonal wind speed for various latitudinal belts over much of the hemisphere, we would find deviations from normal in the

6 Harry Wexler and Namias, "Mean Monthly Isentropic Charts and their Relation to Departures of Summer Rainfall," *Transactions of the American Geophysical Union*, IX (1938), 164–176.

7 L. Teisserenc de Bort, "Etude sur l'hiver de 1879–1890 et recherches sur la position de centres de l'action de l'atmosphere dans les hivers anormaux," *Annales, Bureau Central Meteorologique de France*, IV (1881), 17–62; Carl-Gustav Rossby et al., "Relation between Variations in the Intensity of the Zonal Circulation of the Atmosphere and the Displacements of the Semipermanent Centers of Action," *Journal of Marine Research*, II (1939), 38–55.

Fig. 4 Mean 700 mb Zonal Wind Speed for Western Hemisphere (0° Westward to 180°) for Summer, 1955 Compared to Normal.

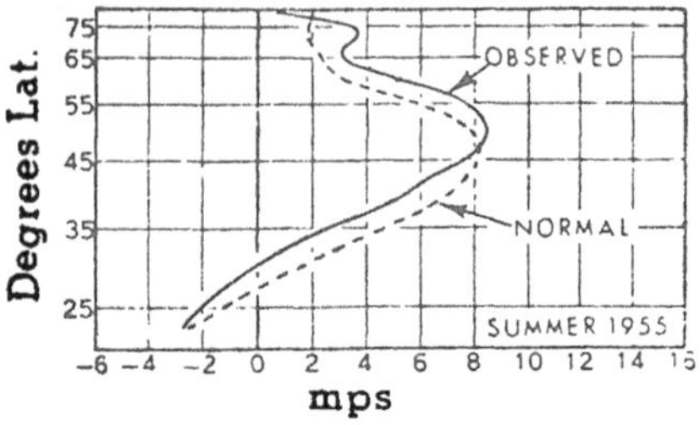

position and strength of the westerlies during droughts. An example is shown in Figures 4 and 5 for the summer of 1955, when the westerlies from 0° westward to 180° were displaced north of their normal position and were stronger in high latitudes than normal. During this summer, the central plains region of the United States was frequently affected by hot, dry conditions, as was northern Europe from Scandinavia through Great Britain.

We can illustrate the teleconnections process with the help of cross-correlation charts as shown in Figure 6. Here we correlate the 700 mb height in the Atlantic high or the Pacific high with all other points 5°'s of latitude and 10°'s of longitude apart. Figure 6 shows that when the Atlantic and Pacific highs are strong during summer, the central United States anticyclone is also likely to be strong.

Whereas the oceanic cells make it more likely for the continental high-pressure cell to emerge and persist, the latter may have its own self-generating properties. Although these properties are not completely understood at present, several theories have been posed to explain this "positive feedback." One explanation is that the land, rendered hot and dry during drought, heats the air above it and further enhances the upper-level isobaric surfaces. A second possibility is that under dry conditions an increase of fine dust particles in the air is likely. These particles would lead to high cloud droplet concentrations whenever cumulus clouds are formed, which could make it more difficult for precipitation to form. A third theory, that has recently received prominence, suggests that high albedo in dry areas (particularly deserts) "contributes a net radiative heat loss relative to its surroundings and

Fig. 5 Mean 700 mb Contours and Height Departures from Normal (both in tens of feet) for Summer, 1955.[a]

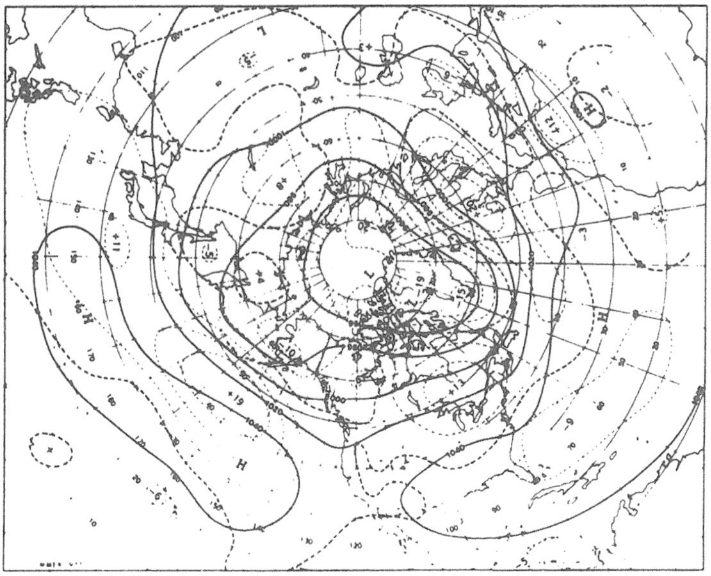

a Note continuous zonal band of positive anomalies in middle latitudes.

that the resultant horizontal temperature gradients induce a frictionally controlled circulation that imports heat aloft and maintains thermal equilibrium through sinking motion and adiabatic compression."[8]

Whatever the mechanisms involved, there is some statistical evidence suggesting that dry, warm springs over the plains of the United States tend to be followed by hot, dry summers. Further, hot, dry summers in the plains have a tendency to persist from one year to the next.[9]

Within the past few years, there have been studies suggesting that the coupled atmosphere-ocean system may account for the

8 S. Twomey and Patrick Squires, "The Influence of Cloud Nucleus Population on the Microstructure and Stability of Convective Clouds," *Tellus,* XI (1959), 408–411; Jule G. Charney, "Dynamics of Deserts and Drought in the Sahel," *Quarterly Journal of the Royal Meteorological Society,* CI (1975), 193–202.

9 Namias, "Factors in the Initiation, Perpetuation and Termination of Drought," in a report issued by International Association of Scientific Hydrology, Commission of Surface Waters, LI (1960), 81–94.

Fig. 6 Teleconnections of 700 mb Heights for a Point in the North Pacific (upper) and in the Atlantic (lower).[a]

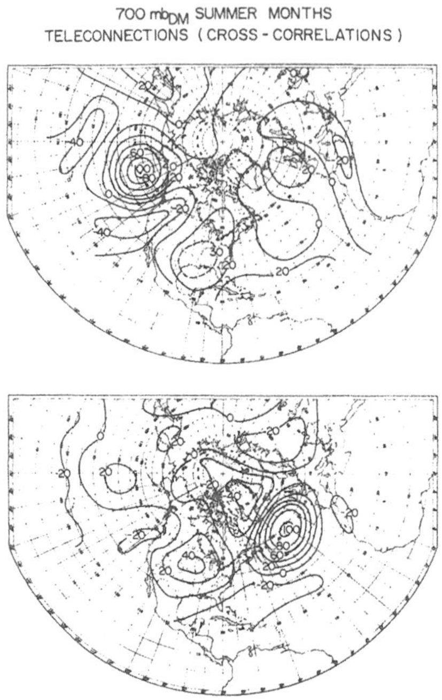

a Teleconnections are cross correlations between the entire field of 700 mb height and the given point. Note in each case the positive correlations over the central United States, indicating the statistical preference for positive height anomalies in the Pacific and Atlantic centers to be aligned with positive anomalies over the central U.S., a drought type situation.

persistence of high-pressure cells over long intervals. As an example, we show in Figure 7 sea-surface-temperature (SST) anomalies associated with droughts over the southern plains region of the United States during the summers of 1952–1954. In each of the summers anomalously warm water was associated with the Pacific and Atlantic anticyclones and consequently the gradients of SST to the north of these warm pools was enhanced. This gradient in SST, transferred to the overlying atmosphere, could have accounted for the strong high-latitude westerlies in these areas. In turn, the high-latitude strong westerlies, if periodically super gradient, could frequently have transferred air southward and thereby could have converged to maintain the high-pressure

Fig. 7 Surface Temperature Anomalies during the Summers of 1952, 1953, and 1954.

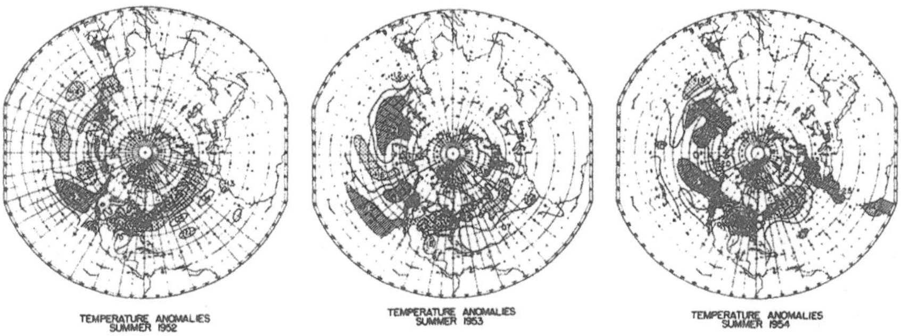

areas to their south in the manner suggested by Rossby. Thus, the clue to predicting mid-continent drought in the United States may lie partly in the coupled air-sea system. Similar comments may apply to the Russian drought in 1972 and to the 1976 drought in Europe and California.[10]

The long waves in the westerlies are known often to occupy favored positions, and theoretical models of the large scale atmospheric flow have demonstrated this. From a climatological standpoint, the east coast of Asia during winter is almost always characterized by a strong trough produced in part by diabatic heating of cold Asiatic air masses moving over the warm Japanese current and by the mountain effects of the Tibetan Plateau. Similarly, the North American Rocky Mountains force a trough to their east through complex dynamic effects on the westerlies. The east coast of North America provides another source for troughs in a mechanism similar to that of the Asiatic trough. Once established, these troughs tend to set up high-pressure ridges downstream at a wave length determined by the strength of the zonal westerlies. The upper wind pattern over the Atlantic and Europe is conditioned in large part by the North American trough.[11]

10 Rossby, "On the Mutual Adjustment of Pressure and Velocity Distributions in Certain Simple Current Systems," *Sears Foundation Journal of Marine Research*, I (1937), 15–28.
11 *Ibid.*; Joseph Smagorinsky, "The Dynamical Influence of Large-scale Heat Sources and Sinks on the Quasi-stationary Mean Motions of the Atmosphere," *Quarterly Journal of the Royal Meteorological Society*, LXXIX (1953), 342–366; Bert Bolin and Charney, "Numerical Tendency Computations from the Barotropic Vorticity Equation," *Tellus*, III (1951), 248–257.

Although these considerations apply to the long-term climatological state, this state is never observed during any one month or season. Since it is the anomalies that are responsible for drought, one must adequately explain the interannual variations and the positions and amplitudes of the long waves. If we could predict a strong trough—let us say in the Central Pacific during winter—we would be in a good position to predict dry conditions along much of the west coast of the United States because of the probability of a strong high-pressure ridge in that area.

Precisely this condition was observed during the strong drought in the winter of 1976–77, as shown in Figure 8. It is possible that the mechanism which maintained the strong trough and responsive west coast ridge was due to the establishment of a strong anomalous gradient of sea-surface-temperature (SST) between the warm Eastern Pacific and the very cold Central and Western Pacific, as indicated in Figure 9. This gradient in SST would not only sharpen atmospheric fronts and troughs, but also produce more southerly wind directions in the zone of temperature contrast. These upper level winds in turn steered cyclones north-

Fig. 8 Winter, 1977 700 mb Contours (solid) and Isopleths of Departure from Normal (broken) both Labeled in Tens of Feet.

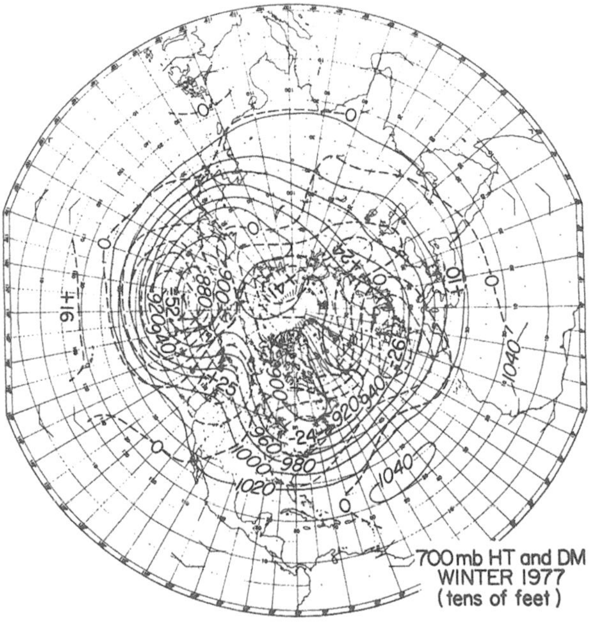

Fig. 9 Sea Surface Temperature Anomalies and Isopleths of 700 mb Height Anomalies in Winter, 1977.[a]

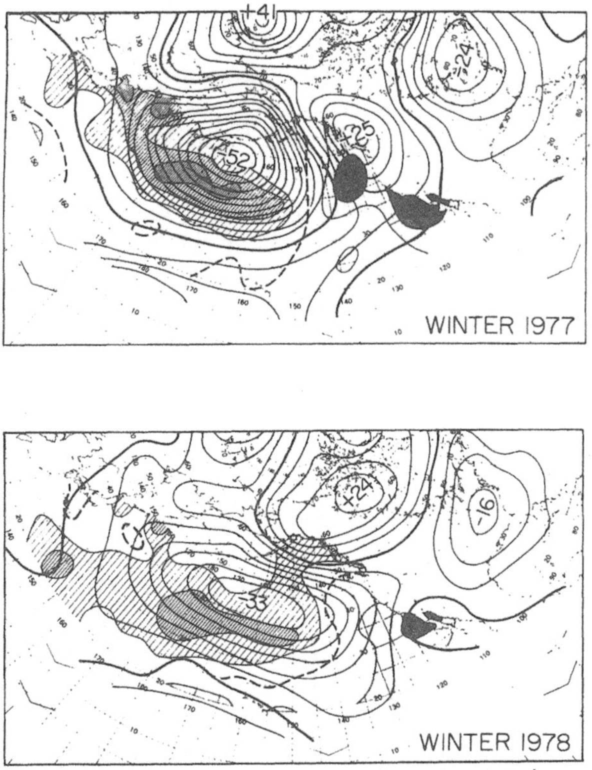

700 mb HT_{DM} ANALYZED TO 50 ft.)
SST_{DM} (ANALYZED TO 1°F)

a Stippling represents sea surface temperature anomalies 1°F or more above normal while shading represents those 1°F or more below normal. 700 mb height anomalies are contoured in intervals of 50 feet.

ward toward Alaska rather than eastward toward the west Coast. In this manner persistent recurrence of these abnormal storm tracks resulted in drought along the west coast. Somewhat similar conditions appear to have been responsible for some of the droughts that occurred in northern Europe during the period 1958–1960, with forcing of the long waves being associated with similar sea-surface-temperature variations in the North Atlantic.[12]

12 Namias, "The Enigma of Drought—a Challenge for Terrestrial and Extra Terrestrial Research," paper presented at the Symposium on Solar Terrestrial Influences on Weather and Climate (Columbus, 1978); *idem*, "Seasonal Persistence and Recurrence of European Blocking during 1958–60," *Tellus*, XVI (1964), 394–407.

Spells of successive years with drought are extremely difficult to explain. It is unlikely that the atmosphere itself has repetitive mechanisms over such long time scales, and it is more likely that persistent anomalous surface conditions or other external forces operate. Proponents of solar-weather relationships believe that the sunspot cycle and accompanying solar variations are responsible for spells of drought. Hypotheses of this kind have appeared in the literature for many decades; the latest efforts in this direction are associated with the twenty-two-year double sunspot cycle. Some meteorologists, however, have been skeptical of solar-weather relationships. If, indeed, solar variations play a part, they complicate the drought problem, because one must forecast not only the atmospheric, oceanic, and cryospheric behavior, but must also know something about the forthcoming sun's activity—a research project of major proportions.[13]

DROUGHTS IN TROPICAL REGIONS The atmospheric features associated with tropical droughts are frequently associated with the position of the intertropical convergence zone (ITCZ), the area where the tradewinds of both hemispheres converge, resulting in ascending motion with copious rainfall. Locations normally occupied by the ITCZ may experience drought if the ITCZ shifts anomalously. There have been suggestions that the position of the ITCZ is contingent upon events in temperate latitudes. Similarly, the El Niño of the equatorial Pacific (a heavy rain producer) has been associated with the trade-wind systems in the equatorial and sub-tropical latitudes of both hemispheres. Although a number of interesting associations have been uncovered, it is fair to say that no one fully understands the reasons for the variations of the ITCZ from one year to another.[14]

13 Evidence for solar-weather relationships has been suggested by J. Murray Mitchell, C. W. Stockton, and D. M. Meko, "Evidence of a Twenty-two year Rhythm of Drought in the Western United States Related to the Hale Solar Cycle since the Seventeenth Century," paper presented at the Symposium on Solar-Terrestrial Influences (Columbus, 1978); Hurd C. Willett, "Solar Climatic Relationships in the Light of Standardized Climatic Data," *Journal of the Atmospheric Sciences,* XXII (1965), 120–136. However, a more pessimistic point of view is found in A. Barrie Pittock, "A Critical Look at Long-term Sun-Weather Relationships," *Reviews of Geophysics and Space Physics,* XVI (1978), 400–420.

14 Namias, "Suggestions for Research Leading to Long-range Precipitation Forecasting for the Tropics," paper presented at the International Tropical Meteorology Meeting (Nairobi, 1974), 141–144; Derek Winstanley, "Recent Rainfall Trends in Africa, the Middle East and India," *Nature,* CCXLIII (1973), 464–465.

CONSEQUENCES OF DROUGHT Prediction of drought has had only limited success; at most modest skill has been achieved in forecasts a month to a season in advance, and longer period forecasting is a current research problem that is largely unsolved. The unpredictable nature of their occurrence and especially their duration has resulted in droughts having tremendous impact on the inhabitants of stricken areas. Statistically, drought may be treated as an extreme event with only a few realizations in the limited sampling domain of observed meteorological history for a given region. Consequently, the probability distribution of drought occurrence is poorly known, and statistical techniques of long-range forecasting of these events are inadequate.[15]

Recent times have illustrated the consequences of drought, with harsh examples scattered across the globe. In the Sahel region of Africa during the late 1960s and early 1970s prolonged drought indirectly killed thousands of people who became weak and susceptible to disease. Livestock herds and crops were destroyed by the drought. The Sahara desert crept to the south, and the defoliated land will take years to recover. In the summer of 1972 temperatures were at an all time high in parts of Europe and Russia. The White Sea (66°N) was as warm as the Black Sea (44°N) and the air temperature climbed to 99°F at Murmansk (68°N), 7° higher than the previous record high. The flow of the Volga River was less than half of normal. In Western Europe in the summer of 1976 temperatures were in the mid-nineties—far above normal. The drought there had severe effects on river and canal traffic as well as agriculture. During the same period, Great Britain's drought was the worst on record since 1727. Vegetable production was down 20 percent, cattle went hungry from lack of feed, and the pound dropped as food prices rose. In the two-year (1976–1977) drought in California, precipitation totaled 55 percent of normal and surface runoff was down to 35 percent of normal. The important agriculture industry suffered greatly from the drought with losses estimated in the hundred millions. Thus,

15 The current status of long-range weather forecasting is treated in Namias, "Geophysical Predictions," *Studies in Geophysics* (Washington, 1978), 103–114.

16 For details of these recent drought episodes, see *UNESCO Courier* (August-September, 1973); Katz, *Unusual Summer*; Evelyn Cox, *The Great Drought of 1976* (London, 1978), 149; California, "The 1976–1977 California Drought."

drought seems to be quite equitable in its appearances, but usually unpredictable in its arrival and duration.[16]

Most cases of drought in temperate latitudes are associated with persistent upper level anticyclonic flow patterns and concomitant subsidence with warm dry conditions in the lower atmosphere. Spatial relationships show that these regional aberrations are really members of an interconnected global family of disturbances, forced by the positioning of the semi-permanent long waves in the atmosphere. The physical processes that conspire to produce these anomalous circulation patterns are only partially understood, but a number of theories suggest that influences exterior to the atmosphere are involved. Among these, boundary conditions affecting the surface heat exchange such as sea-surface temperature anomalies and surface albedo changes have been advocated. Changes in the solar radiation input, as indicated by sun spot activity, have also been said to influence long period changes.

David Herlihy

Climate and Documentary Sources: A Comment

In the investigation of climates which have prevailed in historical times, documentary records offer the researcher certain distinct advantages. At least in some parts of the world (notably in Western Europe), written documents affording some information on climate exist in continuous series, stretching back into the Middle Ages. They are, for the most part, readily available in publications and archives; they are easily read and cheaply processed. But their use and interpretation also present the researcher with certain formidable difficulties.

As Christian Pfister has lucidly explained, our knowledge of past climates is based on two groups or families of data. The one he calls "field data"; it consists of observations taken upon natural materials, usually strata laid down over time and influenced in their formation by the then prevailing climate. The second group, which he calls "documentary data," consists of written records.

Whether taken from field or from document, all usable data concerning historic climates must clearly satisfy two conditions. The individual observations must be accurately datable, or, in another phrase, the time resolution of the series must be high. The researcher needs to locate the historical "moment" to which the observation refers. Moreover, within the same series, the observations must be precise, consistent, and comparable, and thus capable of supporting sophisticated mathematical or statistical analysis.

Here, however, a fundamental difference between our two families of data becomes apparent. Each family readily satisfies one, but not the other, of the two conditions. Field data, on the one hand, examined by rigorous physical, chemical, or radiological methods, can yield highly precise and fully comparable measurements. But the individual observations in field data are often difficult to date accurately. Even an approximate dating of plus or minus fifty years, although marvelously precise on the scale of geological time, does not satisfy the needs of most his-

David Herlihy is the Henry Charles Lea Professor of History at Harvard University. He is the author, with Christiane Klapisch, of *Les Toscans et leurs familles* (Paris, 1978).

torians. To be sure, some types of field data—tree rings, for example—allow higher time resolutions, but the dating of a tree ring still requires considerable study and expense. Chronological precision may be possible, but it is not easily achieved.

Documentary data, on the other hand, show exactly opposite qualities. Usually, the individual observations forming part of a time series can be exactly dated, by day and month as well as by year. But before the development of scientific instrumentation, the observations lacked adequate precision, and this weakens both their comparability and the power of the mathematical analysis which they can bear. In sum, neither one of these two families of data offers an entirely satisfactory record of the climates which have prevailed in historical time.

At first brush, these defects might seem self-correcting. Could not the high time resolution of the documentary series be combined with the precise observations of field data, in order to develop a fully satisfactory record of past climatic change? Associations of this kind are frequently made in the literature. Pollen deposits in an earth core are read in the light of the presumed settlement history of the region; stratigraphical evidence of various sorts is linked with the little ice age of the late sixteenth century, a phenomenon which was itself discovered through historical documents. Still, on the most fundamental level, this remains questionable procedure. There seems no way of establishing the certitude of such associations, however likely they may appear. A risk of false associations is consequently present, which may offer specious conformation to mistaken judgments. This is not to say that the valid conclusions drawn from both sets of data ought not to be combined in a larger picture of past climates. But the conclusions have first to be established independently, by methods appropriate to the data; otherwise they cannot command complete confidence.

The documentary evidence—our chief interest here—can further be classified into two subsets. The documents may record direct observations of meteorological events, or they may describe weather-related phenomena, such as the abundance of harvests, the movement of food prices, and the like. Observations of this latter kind are called, in the now standard phrase, "proxy data."

Helmut E. Landsberg's article, "Past Climates from Unexploited Written Sources," provides an eminently informed survey of surviving weather observations made chiefly in northern Europe during the early modern period. With an admirable command of his materials, he notes the difficulties in the data, stemming from shifting or uncertain metrologies, limited number of observational stations, interrupted series, and the like. Still, the variety and age of these observations are impressive. Landsberg also notes that this series of direct observations could be enlarged through the systematic exploration of diaries and journals. I would add that even chronicles and annals contain frequent allusions to meteorological events. For Europe, at least, it ought to be possible to compile a kind of almanac of meteorological occurrences and conditions, beginning as early as the central Middle Ages.

The deficiencies in such a historical almanac are also readily apparent. Observers were most impressed by unusual happenings and by extreme conditions. As Jan de Vries correctly observes, this sort of weather chronicle presents a record of discrete, often disastrous events, and tells us little about the abiding conditions and lasting influences of climate in the past. Moreover, the observers differed one from the others in culture and curiosity, and this subjectivity severely weakens, if it does not entirely destroy, the comparability of the data over time. And yet this almanac of observations has value, preeminently for two reasons. It is useful to know the extreme conditions and dramatic events which prevailed or happened in the past, impressed contemporaries, and thus became part of recorded history. And for historians at least, it is valuable to have a chronicle of human perceptions of and reactions to weather; attitudes toward the natural world form a principal part of the collective mentality and culture of historical societies.

The use of proxy data in the examination of historical climates presents special difficulties, which de Vries and Andrew B. Appleby forcefully describe. And yet the substantial advantages of proxy data should be noted too. Proxy data are, to begin with, much more abundant in the historical record than direct observations of meteorological phenomena. Many European archives, for example, contain series of food prices, reflecting the abundance of harvests at least indirectly; some of them begin already from the late thirteenth century. And the proxy data, such as prices, are often in numerical form; they support much more readily than descriptive observations comparisons over time and precise analysis.

One type of proxy data which appears particularly promising is "phenological" data, based on the development or maturation of plants. Emmanuel Le Roy Ladurie and Pfister make use of the dates of the grape harvests, in early-modern France and Switzerland, as evidence of the changing climate. The ingenuity of the approach is manifest, but here too a word of caution is required. Researchers exploiting records of this kind must still apply to them the basic questions of historical criticism: why was this information recorded? and how did the purposes and manner of its collection affect its character and accuracy? At least in the articles here reviewed, the authors do not adequately inform their readers concerning the exact nature of the data that they are using.

Why were the town governments of France and Switzerland interested in the maturation of the grape? And what exactly do these dates mean? The dates were not a simple and straightforward observation on the maturity of grapes, but formed part of a governmental decision, an order, that growers were not to begin harvesting the grapes before the stipulated time. This assured that no grower within the jurisdiction would gain a competitive advantage by harvesting the grapes in advance of his neighbors. The setting of the grape harvest thus represents the characteristic effort of governments under the old regime to regulate economic enterprise, in an effort to suppress allegedly unfair competition.

The selection of the date at which the harvest could legally begin was thus an administrative and political decision, and was inevitably subject to some political influences. And in the selection of a common date, it was not always easy to reconcile the interests of grape growers on highlands or lowlands, or of big and small producers. The decision at times evoked resentment and protests. In the Italian city of Pistoia, for example, in the middle fourteenth century, the government tried for some ten years to impose a common date for the beginning of the grape harvest, but encountered such strong resistance in the countryside that it abandoned the policy. Probably for this reason, the Italian city-states, in spite of their flourishing bureaucracies, rarely attempted to set legal dates for the initiation of the grape harvest.

Researchers, in sum, need to recognize and examine the political factors influencing the dates of the harvest. It may be that tendencies toward greater economic liberalism or toward more stringent controls affected the decision. Even changes in the extent of land given over to vineyards could have had an influence, if, for example, a larger area of highlands, where grapes matured slowly, was planted in vines. Most likely, the value of the series for climatological research will survive critical scrutiny; but the scrutiny itself remains indispensable.

The dates of the grape harvest, although they must be used critically, are an example of a particular type of proxy data which seems especially promising for climatological research. Traditional societies did not measure many things well, but they counted accurately enough the passage of days, months, and years. The stated dates allow, as we have mentioned, the observations to be set precisely within a chronological series. Further, the dates can occasionally serve another function: they can be used as parts of the observations themselves. Pfister, in his admirably comprehensive survey of Swiss weather from 1525 to 1825, gives several examples of this strategy, and many more could be cited: the study of dates when particular rivers or harbors froze or thawed, when cherry trees blossomed in medieval Japan, and the like.[1]

It seems to me that this promising method, exploiting one of the unique aspects of the documentary record, could be extended into still other areas. Many European archives, for example, have preserved weekly lists of food prices, which sometimes identify local and imported produce. Food prices in turn rose and fell in yearly cycles directly linked to the harvest. By studying the yearly cycle of locally produced food, and by identifying departures from the normal or average cycle, it might be possible to link these fluctuations with seasonal weather. The investment of labor would be large, but it would doubtlessly also provide the historian with much useful information on the operation of seasonal markets.

The use of written records in paleoclimatology confronts the researcher with formidable obstacles, but there are grounds for moderate optimism. We have not systematically utilized all the extant documents. Then too, as the foregoing articles show, we have not exhausted our ingenuity in devising new ways of using them.

1 This strategy is also employed in Pfister, "Climate and Economy in Eighteenth-Century Switzerland," *Journal of Interdisciplinary History*, IX (1978), 223–243.

John D. Post

The Impact of Climate on Political, Social, and Economic Change: A Comment

That climate changes is well established. At issue is whether climate change has had any significant influence on the fluctuation of demographic variables or the level of economic activity, beyond its impact on those lands located at the margins of the temperate zone.

According to Lamb and others, the preindustrial centuries of Europe, the period discussed in the papers of Andrew B. Appleby and Jan de Vries, were marked by low values of the long-term mean temperature in comparison with recent normals. This little ice age, stretching roughly from the last half of the sixteenth to the first half of the nineteenth century (with several warm interludes), was also characterized by enhanced variability of temperature from year to year and from decade to decade, particularly between 1550 and 1700. In some decades (the 1570s, 1590s, and 1620s), wet, cool summers constituted the most notable anomaly; in others (the 1560s and 1600s), it was the severity of the winters; whereas in the 1580s, 1680s, and 1690s, both summers and winters proved cold.[1]

Presumably, the relatively high number of anomalous decades reflects the enhanced climatic variability of the little ice age. At the same time the long-term negative temperature departures from recent normals seem modest, some 1°C in global means. Although a change of this magnitude would produce significant ecological consequences on a secular time-scale, when we deal with human ecology attention must be given not only to population and the natural environment but also to technology, which has enabled human societies to buffer the influence of weather with notable success.

The studies of Appleby and de Vries cast serious doubt on the impact of climate change on mortality levels and economic activity. In part, however, such findings derive from England and Holland, where the levels of economic development and tech-

John D. Post is Associate Professor of History at Northeastern University. He is author of *The Last Great Subsistence Crisis in the Western World* (Baltimore, 1977).

1 Hubert H. Lamb, *Climate: Present, Past and Future* (New York, 1978), II; Hans von Rudloff, *Die Schwankungen und Pendelungen des Klimas in Europa . . .* (Braunschweig, 1967).

nology provided insulation against meteorological shocks beyond that of any other region of preindustrial Europe. To obtain different results, we would not have to turn, as Appleby suggests, to marginal locations such as Scandinavia and Scotland, but simply to any other region of Europe, including the relatively developed French and German speaking areas. Nonetheless, the conclusions reached are substantially valid for England and Holland, even if they may turn out to be progressively less applicable as one moves down the ladder of economic development.

In Appleby's article, his finding that neither the specific epidemiology of such lethal contagions as plague, smallpox, and malaria, nor their statistical correlation with weather patterns, discloses any significant dependence on temperature values or humidity levels uncommon to Europe, is valid. At the same time, anomalous weather conditions have produced subsistence crises and mortality peaks resulting from both starvation and infectious disease. The question is whether periods of meteorological stress promote epidemic disease. Appleby's study finds little connection between high food prices and the incidence of infectious disease.

However, in the European subsistence crises of 1740–1741 and 1816–1817, the major health problems were traceable to both weather conditions and high grain prices. The higher mortality rates resulted mainly from epidemics of dysentery-diarrhea (the two cannot be distinguished in the absence of laboratory tests) and the louse-borne infections of typhus and relapsing fever. These epidemics were induced by a combination of poor weather (cold winters, wet springs and autumns, and dry summers) and the vagrancy and begging promoted by high food prices and unemployment.

How do we account for the fact that Appleby's analysis of the London Bills of Mortality and English parish registers failed to reveal the elevated incidence of these infections? First, the London Bills at the time of these two crises did not distinguish between fever and typhus; and thus the epidemics must be documented from medical clinical evidence, which is unambiguous in identifying both typhus and relapsing fever. Second, with respect to dysentery-diarrhea, included in part in Appleby's study under infantile mortality from convulsions, this major cause of death was primarily a rural phenomenon; by contrast, the majority of available time series of burials and deaths derive from urban locations. Dysentery was known almost everywhere in Europe as the "country disease," and Europe's preindustrial population was more than 80 percent rural. These facts by no means demonstrate that the little ice age induced a higher incidence of epidemic disease, but they do indicate that periods of marked climatic variability foster some epidemic diseases.

De Vries' finding of no significant correlation between climate change and the level of economic activity is convincing, at least to the degree that the Dutch experience was representative of Europe's preindustrial economy, and to the extent that the meteorological periodization utilized in the statistical tests reflects the critical fluctuations. The Dutch economy was the most insulated against such external shocks as climate change or variance. As de Vries has concluded elsewhere, the Dutch economy in the seventeenth century "seems to have been in the unique position of being a net gainer from the events that brought grief to much of the rest of Europe." This circumstance was traceable in part to Holland's commanding position in the international grain trade. The Dutch economy was shaped the least by subsistence cereal cultivation; real income probably declined least in Holland during European subsistence crises.[2]

De Vries is aware that correlating the annual variance of winter temperature and the annual fluctuations of such economic variables as rye prices will not necessarily provide adequate answers; instead, recourse must be had to a more elaborate model. For example, the relationship between the growth of winter grains and the weather patterns in the temperate zone includes eight separate weather phases that extend over the entire twelve-month period. Christian Pfister has found that in the Swiss lowlands heavy autumn rainfall was primarily responsible for the poor winter-grain harvests by preventing the peasants from plowing early enough in the fall and thereby reducing the area sown. De Vries has offered evidence demonstrating not only that rainy winters affected some Dutch arable farming but that farmers were able to take steps to weaken the causal link by drainage technology. But it is doubtful that this adjustment to anomalous precipitation could be generalized throughout the European grainlands of the eighteenth century.

2 Jan de Vries, "Barges and Capitalism: Passenger Transportation in the Dutch Economy, 1631–1839," *A. A. G. Bijdragen*, XXI (1978), 325.

The great drama, as de Vries has put it, resides in the question of whether climate change has impinged upon human history. His skepticism is well founded, for the long-term consequences of climate change have not been convincingly demonstrated. But neither has the issue been decisively resolved. In demographic history, for example, it is well established that the population of Europe grew only slowly during the seventeenth and eighteenth centuries (except in the Netherlands and England). This was striking when contrasted with the rapid population growth during the twelfth and thirteenth centuries, when the European climate was apparently milder and less variable. We still do not know what role, if any, the severe cold phase of the little ice age played in this demographic stagnation. We do know that Europe's preindustrial population growth was slow despite the wide diffusion of new or more intensively cultivated crops such as buckwheat, barley, oats, rice, maize, and potatoes, which should have buffered the impact of anomalous weather patterns on total crop yields; by contrast, despite the almost exclusive reliance on winter grains, the centuries of the medieval climatic optimum witnessed a pan-European population explosion. Climate change could not have been the only major variable involved; if nothing else, as Appleby noted, comparative studies demonstrate that some European regions have been less influenced by climate than others.

De Vries' emphasis on the consequences of climate change that flow from differences in variance is well placed. Long-term changes in the variance of climatic phenomena probably did elicit adjustments in the technologies used by farmers. On the other hand, the severe winter of 1740 is singled out as a case when climate variance produced both economic and demographic effects. That winter had been preceded by a quarter century of milder "maritime" climate with mostly moderate winters and, on balance, reduced variability. Since technologies had evolved based on the perceived climate probabilities, they were able to buffer only imperfectly the dislocations created by a severe winter.

This hypothesis is no doubt valid; but an empirical examination of the weather patterns of the 1730s and 1740s discloses a need for some amplification. To begin with, the severe winter of 1740 was not only among the coldest in preindustrial Europe; it proved the longest both in terms of the duration of subnormal temperatures (approximately from the previous September to the following May) and the number of days exhibiting frost. Moreover, the weather patterns continued anomalous for the remainder of the year (drought in the summer, wet and cold in the autumn, and excessive precipitation and floods in the winter months). In addition, the year 1740 followed several years marked by poor weather during the growing seasons and elevated grain prices; 1740 was in turn followed by several years marked by cold winters and drought-ridden summers. In spite of the accumulation of meteorological stress, the Dutch Republic experienced modest economic dislocations and minor demographic consequences in comparison to the remainder of Western and Central Europe, which suffered a mortality peak from 1740 to 1743.

The foregoing discussion suggests that neither secular climate change nor climate variability (i.e. fluctuations over a year or two) may be the key interval for the study of the possible impact of climate on mortality rates and economic activity. The decade or a series of years longer than the term climate variability implies may represent the appropriate historical periodization. Human societies seem to be able to accommodate a temperature fluctuation of 1°C, which is the order of magnitude involved in secular climate change. With regard to climate variability, European preindustrial societies developed a large measure of resilience based on the insulation provided by better crop mixes, emergency grain stocks, an expanding international grain trade, and an increasingly effective welfare system. Probably only after a succession of severe, bad, and poor weather years would this system of adaptations and adjustments be overchallenged, leading to higher death rates and some decline in economic activity. If research discloses that even a decade of anomalous weather conditions, occurrences which were more frequent in preindustrial Europe, failed to produce significant demographic and economic effects, then we can conclude that the direct human consequences of climate change have been slight.

John A. Eddy

Climate and the Role of the Sun

> It is true that from the highest point of view the sun is only one of a multitude—a single star among millions—thousands of which, most likely, exceed him in brightness, magnitude, and power. He is only a private in the host of heaven. But he alone, among the countless myriads, is near enough to affect terrestrial affairs in any sensible degree; and his influence upon them is such that it is hard to find the word to name it; it is more than mere control and dominance.[1]

OUR NEED FOR SOLAR CONSTANCY Our utter dependence on the sun for daily light and heat is so obvious a fact of our existence that it is easily overlooked. To this one star we also owe all our food—through a chain of life that begins in simple plants and aquatic forms—the replenishment of oxygen through photosynthesis, and the generation of nearly all of the energy that we have ever used. In harnessing wind and water power we reap the solar energy that drives the atmospheric circulation and cycles the water from ocean to air and back again, and in burning wood, coal, or petroleum we are harvesting the sunlight that fell on the earth in the past.

The same logic leads us to look into the possible variability of the sun, and to examine the degree of our susceptibility to known or suspected solar variations. How constant is the sun, and how dependent are we upon its constancy? Our fate would be dire, indeed, were the sun to go out altogether, or loom to nova brightness. But what of more astronomically probable changes, say of 10 percent, or 1 percent in total solar flux? What might we feel from unseen and less energetic changes in the sun's ultraviolet radiation, or in its output of atomic particles and magnetic fields? Could it be that terrestrial weather is controlled by variations on the sun?

John A. Eddy is Senior Scientist at the High Altitude Observatory, Boulder, Colorado.

The work on which this article is based was sponsored by the Langley-Abbot Program and the Scholarly Studies Program of the Smithsonian Institution, and by NASA Grant NSG-5345.

1 Charles A. Young, *The Sun* (New York, 1896), 1.

The sun is the engine that drives the atmosphere; even minor changes in its output could alter atmospheric composition, temperature, or circulation. These changes, if persistent, could influence the long-term average of weather—called climate—and through climatic change bend the path of human progress. Present mathematical models of global climate suggest that a decrease as small as 1 percent in the total radiation output of the sun is adequate to lower the global temperature average by 1 or 2°C, and hence to bring about a little ice age, of the sort that gripped Europe and America throughout the seventeenth and eighteenth centuries. The same numerical simulations indicate that a change of only 0.1 percent in solar flux, if continued long enough, could bring about climatic changes of significant social and economic impact. And a change of 10 percent, in a negative direction, would be globally disastrous, inducing major glaciation and perhaps an ice-covered earth which could recover only after a far greater and less likely increase of about 50 percent.[2]

To date, however, nothing significant that has ever been recorded in the meteorology of the lower atmosphere of the earth *requires* any change in solar output for its explanation. The weather system, or our present limited understanding of it, could work well with a star of perfect constancy. The succession of day and night, the march of the seasons, and even the recurrence of the awesome ice ages are all nicely explained by the movements and orientation of the earth itself, with no need for any solar variability at all. Shorter excursions of climate and the vagaries of daily, local weather can be explained, if need be, by random variations or by any one of several competing mechanisms that are internal to the atmosphere itself.[3]

Thus the question of the role of the sun in bringing about changes in weather or in climate turns back to whether it is, or is not, a variable star.

SUNSPOTS AND THE SOLAR CYCLE We have long known that the sun varies. With the first telescopes, early in the seventeenth

2 Hubert H. Lamb, *Climate. Past, Present, and Future* (London, 1972); W. L. Gates and Y. Mintz, *Understanding Climate Change* (Washington, D.C., 1975), 149–152; Emmanuel Le Roy Ladurie (trans. Barbara Bray), *Times of Feast, Times of Famine: History of Climate Since the Year 1000* (Garden City, 1971), 129–226.

3 James D. Hays, J. Imbrie, and N. J. Shackleton, "Variations in the Earth's Orbit: Pacemaker of the Ice Ages," *Science*, CXCIV (1976), 1121–1132.

Fig. 1 A Seventeenth-Century Drawing of the Sun and Sunspots.

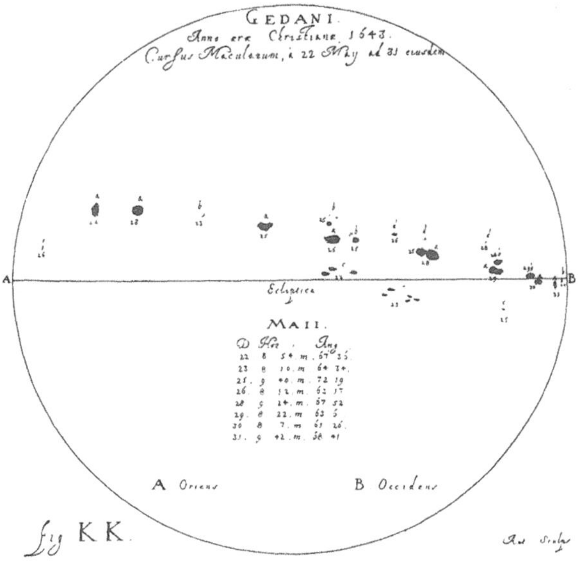

From Johannes Hevelius, *Selenographia sive Lunae Descripto* (Danzig, 1647).

century, came the realization that the sun, like all else in nature, was ever changing. Small dark spots were seen on its surface, and their number and positions changed from day to day (Fig. 1). Large sunspots can be detected with unaided eye under favorable observing conditions; descriptions of them can be found as early as 28 B.C. in records from the Orient. But it took the telescope to reveal their varied forms and regular motion, and to establish that they were indeed features of the sun itself and not intervening planets or objects in our own atmosphere.[4]

The concept of a blemished sun was at first resisted in Western thought and theology, although soon accommodated by the rationalization that the dark spots were only clouds that moved across the face of an otherwise perfect sun. Today we know that sunspots are deeply rooted disturbances; they mark the places where intense magnetic fields break through the solar surface. These varying magnetic fields produce other dynamic changes on the sun and identify it as a magnetically variable star. Although sunspots look small, their average size is about that of the earth

4 David H. Clark and F. Richard Stephenson, "An Interpretation of Pre-Telescopic Sunspot Records from the Orient," *Quarterly Journal of the Royal Astronomical Society,* XIX (1978), 387–410.

and the patterns of magnetic field with which they are associated cover all of the sun.

Since the middle of the last century we have known that the sun produces sunspots in a fairly regular cycle of about eleven years, during which time their number, counted on any day, rises from none, or almost none, to as many as several hundred, and then back to zero again. Many other features of solar variability follow this same rule, and thus the eleven-year sunspot cycle has come to be synonymous with solar variability and is probably the best known feature of the sun.

The physical reason for the cyclical production of sunspots is still not clearly understood. Many characteristics of the cycle can be reproduced in numerical models of the solar atmosphere as a product of the interaction of magnetic fields within the sun and the observed properties of solar surface rotation, through a mechanism known as the solar dynamo. Neither models nor measurements give reason to suspect that the total luminosity of the sun varies significantly with the solar cycle.[5]

SUNSPOTS AND WEATHER Far more effort has been expended on the search for terrestrial implications of sunspots and the sunspot cycle than has ever been spent on their physical explanation. Much of this work has been highly speculative and subjective. Indeed, since their earliest discovery, sunspots have often been associated with mysticism and things occult. They were taken as omens by the Oriental astrologers who first recorded their appearance. Later telescopic identification was accompanied by only slightly more objective efforts to trace their effects on the earth, chiefly through their presumed diminution of sunlight.[6]

Such speculations were common enough in the seventeenth and eighteenth centuries to be found in literary references. Andrew Marvell made allusion to the deleterious effects of sunspots on solar light in a poem written in 1667, as did John Milton, less directly, in *Paradise Lost* when he described the allegorical landing of Satan on the sun as like a sunspot seen through a "Glaz'd Optik

5 Michael Stix, "Dynamo Theory and the Solar Cycle," in V. Bumba and J. Kleczek (eds.), *Basic Mechanisms of Solar Activity* (Dordrecht, 1976), 367–388.
6 Clark and Stephenson, "Sunspot Records," 402.

Tube," or telescope. In *Gulliver's Travels,* written in 1723, Jonathan Swift included the fear of sunspots and their dimming of the sun among the grave celestial worries of the Laputan astronomers.[7]

In 1801 William Herschel, the renowned astronomer, attempted to quantify the association with a cautious correlation that he had found between times of unusual sunspot absence and the English weather, as reflected in historical records of the price of grain on the London market. Thus the search was launched long before the cyclical nature of sunspots was found. But it was the demonstration of the cyclical regularity of spots, with the work of Heinrich Schwabe in 1843, that sparked explosive interest in the subject. Cyclical phenomena have a hypnotic attraction, particularly felt at the fringes of science, and the apparent cosmic regularity and presumed significance of the sunspot cycle has drawn a steady stream of determined attempts to link its ups and downs with searched-for periods in weather, agriculture, economics, health, and human behavior.[8]

From the start these attempts have focused on a possible connection with weather. These efforts, by scientists and by laymen, began in earnest in the middle of the last century. The reason, surely then as now, lay not so much in scientific logic, for there were no measurements of periodic solar outputs and no previously recognized cycles of eleven years in weather records. Then, as now, the search was based instead on the hope of finding a key to practical weather prediction. If a significant weather connection could be found with what then seemed to be a regular, solar clock, the gross nature of climate variation could be predicted—a year, ten years, or 100 years in advance. Thus the original motivation was not *a priori* indications of a sun-weather

7 The poem is Marvell's *The Last Instructions to a Painter,* written in 1667. It is discussed in Judith E. Weiss and Nigel O. Weiss, "Andrew Marvell and the Maunder Minimum," *Quarterly Journal of the Royal Astronomical Society,* XX (1979), 115–118.

8 William Herschel, "Observations Tending to Investigate the Nature of the Sun . . . ," *Philosophical Transactions of the Royal Society of London,* XCI (1801), 265–318; Heinrich Schwabe, "Die Sonne," *Astronomische Nachrichten,* XX (1843), 282–286. Examples of such studies can be found in Harlan T. Stetson, *Sunspots and Their Effects* (New York, 1937) or, more recently, in M. N. Gnevyshev and A. I. Ol' (eds.), *Effects of Solar Activity on the Earth's Atmosphere and Biosphere* (Jerusalem, 1977) which I have reviewed in *Icarus,* XXXVII (1979), 476–477.

connection, or any solid rationale that there should be one, but rather the common good that would accrue if one were found.

There is nothing wrong with this approach. It was the logic often used, for example, by Thomas Edison in tackling technical problems, sometimes successfully, often not. But such an approach, when applied to investigative science, the ultimate goal of which is to understand a larger system (in this case, the atmosphere and weather) can be a crippling constraint.[9]

At first it seemed to work. The 1870s and 1880s were characterized by a flurry of scientific papers purporting to have found connections between solar behavior, and more particularly the eleven-year sunspot cycle, and the weather—as measured in monsoons in India, rainfall in Ceylon, temperature in Scotland, or the depth of the rivers Thames, Elbe, or Nile. A new day had dawned, some said, in which science had made possible the prediction of the future, bestowing an ability to anticipate and thus to conquer problems that had long plagued mankind. "The riddle of the probable times of occurrence of Indian Famines," announced Sir Norman Lockyer in 1900, "has now been read, and they can be for the future accurately predicted." That prediction, we need hardly explain, lay in a pattern of recurrence that, according to Lockyer's own researches, seemed to fit the ups and downs of the sunspot cycle.[10]

And then the bubble burst. One by one, the simple relationships vanished when examined more critically or faded in the light of longer records. By and large scientists have come to recognize these early, naive relationships as accidental coincidences in limited data sets—examples of what Langmuir would later call "pathological science," in which our own desire to find a certain result influences what we see or do not see.[11]

Newer knowledge of both sun and weather make the simple cause-and-effect chains unlikely, or, if at work at all, more likely masked by a myriad of other, more energetic changes in the atmosphere. Moreover, modern measurements of the sun's out-

9 Matthew Josephson, *Edison* (New York, 1959); Eddy, "Edison the Scientist," *Applied Optics*, XVIII (1979), 3736–3750.

10 A. J. Meadows, *Science and Controversy* (Cambridge, Mass., 1972), 124–127.

11 Irving Langmuir (ed. R. N. Hall), "Pathological Science," General Electric Company Report 68-C-035 (Schenectady, 1968).

puts have as yet failed to detect any unequivocal change in the total solar radiation with day to day changes on the sun, or with the sunspot cycle.[12]

We do know that the ultraviolet and x-ray radiations from the sun vary appreciably with solar activity, and that the sun's flow of atomic particles and magnetic fields are constantly changing. But these are all minor perturbations in the total budget of solar energy that we receive, and in each case they affect the earth directly only at the topmost layers of the atmosphere. If these upper atmosphere effects are to influence the denser troposphere, far below, we need to find a way for the tail to wag the dog. It is still possible that these subtle changes in some way make their mark on weather, and so the search goes on. But it is a search that now looks not so much for practically important connections as for minor perturbations through far more subtle chains.[13]

LONG-TERM SOLAR VARIABILITY It is possible that we have missed the forest for the trees. Driven by pragmatic hopes of finding keys to weather prediction, we run the risk of concentrating too much on time scales of more practical consequence—of days, months, or years. In taking a longer view we see the problem in clearer perspective. We may also expect to identify extremes of sun or weather behavior that can serve as more sensitive tests.

Physics and logic are here on our side. If there exist immediate sun-weather connections that hide in the noise of random weather variations or are masked by patterns of stronger, competing effects, we can hope that they would become more visible and better defined when solar behavior reaches long-term extremes. Moreover, because of the physical inertia of the earth's atmosphere, we expect persistence to win out over impulse; the longer a solar effect is applied, the more likely it is to exert a real influence on the atmosphere. If solar behavior changes consistently for long enough we can expect to find the mark of even as

12 Claus Frohlich, "Contemporary Measures of the Solar Constant," in Oran R. White (ed.), *The Solar Output and Its Variation* (Boulder, 1977), 93–109; Peter V. Foukal, P. E. Mack, and J. E. Vernazza, "The Effect of Sunspots on the Solar Constant," *Astrophysical Journal,* CCXV (1977), 952–959.

13 White (ed.), *Solar Output,* 131–348, 349–403.

subtle an effect as a recognizable signature in climate records. The vast scale of the sun and the appreciable depth of the solar convective zone argue as well for slower changes, moderated by thermal and mechanical inertia that put constraints on allowed variations in the sun's radiative output of energy. These constraints relax as we consider longer and longer time periods.

The sun is now thought to be about 5 billion years old. During this fraction of its expected life it has evolved considerably and, since stellar evolution is a continuous process, there is no reason to think that in our present era the sun rests, as on a cosmic Sabbath. We have often erred when we presume our present time or place in any way unusual.

From geological and paleontological evidence we can put coarse limits on the range of possible past excursions in the output of solar radiation. Because we find in rocks primitive forms of life that are 2 billion years old, we conclude that in subsequent time the earth has probably not been so hot that all the water boiled or so cold that all the oceans froze. These considerations allow us to put limits, using present climate models, of about ±10 percent on changes in total solar radiation, assuming that the composition of the atmosphere has remained constant during this time.[14]

Between these inferences on geological time scales and the hard evidence from the modern era of detailed observations of the solar surface, there are both time and opportunity for much to have happened. Our intensive observations of the sun span at most a century—less than .00001 percent of the time that it has thrown its light and heat upon the earth. Is this brief sample typical? So small a fraction is surely inadequate to tell us all the history of the sun. It is presumptuous to suppose that we have been so lucky to have seen, in the wink of time that we have watched the sun, all of the changes that it knows, or that the spectrum of solar behavior allows evolutionary changes of billions of years, and immediate changes of eleven years or less, but nothing in between.

The problem is that solar history, like all history, is imper-

14 Hays, "Climatic Change and the Possible Influence of Variations in Solar Input," in White, *Solar Output,* 73–90.

fectly known. In taking the longer view we see farther but less distinctly, and look through mists of time that soon close in, leaving us with dim generalizations derived from facts inferred or indirectly obtained. The sun appears early in history and even pre-history as a common symbol, an object of worship, and a device to tell the time of day or year. But accounts of its physical appearance, which we need to describe the sun's behavior as a star, do not begin until late in human history.

The earliest known of these direct physical observations tell of sunspots seen with naked eye in the first century B.C., in China, giving us a potential record of 2,000 years. Uncertainties in all the naked eye accounts, and their sporadic nature, severely limit their utility. We generally speak of reliable solar history as beginning with the telescope, in 1610.

THE LIMITS OF DIRECT HISTORICAL SOLAR OBSERVATIONS Our oldest direct record of solar behavior is the relative sunspot number, a measure of the number of spots seen on the sun with simple telescopes in unfiltered light. It is most often presented, as in Figure 2, as an annual average of daily measurements made at cooperating observatories over the earth. The sunspot number has been determined from direct observations in this way since 1848, when the quantity shown was first defined. Earlier values, shown to 1610 in Figure 2, were reconstructed from historical data and are much less reliable. The quality of the historically derived record is itself far from uniform, for the older records are successively poorer, particularly with respect to amplitudes of the peaks. This lack of uniformity is not because the older observers were less able, or less careful, but that the records are less continuous, making annual averages less meaningful since the observed sunspot number varies appreciably from day to day and is strongly modulated by solar rotation. Moreover, the sunspot number, as defined, includes a subjectively determined correction factor to account for sky conditions, the telescope used, and the idiosyncrasies of the observers. These factors are at best poorly recovered from historical data.[15]

The sunspot number describes solar activity as it is seen in

15 Max Waldmeier, *The Sunspot Activity in the Years 1610–1960* (Zurich, 1961).

Fig. 2 Annual Mean Sunspot Number, A.D. 1610–1975.

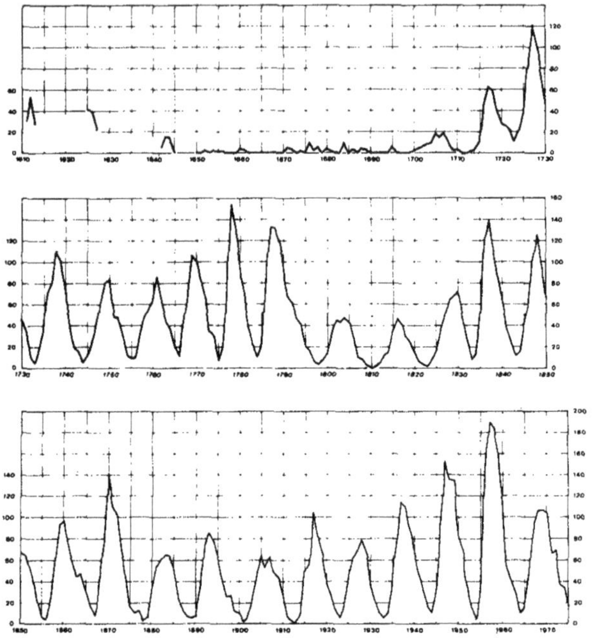

SOURCES: Waldmeier, *Sunspot Activity*, 110; Eddy, "Maunder Minimum."

a single, shallow layer of the solar atmosphere called the photosphere—the part of the sun most easily observed and from which most of its light and heat emanate. Modern solar observations of higher regions of the solar atmosphere, made with monochromatic filters in the visible spectrum and by other techniques in ultraviolet, x-ray, and radio wavelengths, have shown that the layer in which sunspots appear is deceptively quiet. Thus the sunspot number is not an especially sensitive indicator of solar activity or solar change.

In an analogy with the stock market, the sunspot number is a kind of subjectively defined Dow-Jones average that serves as a general indicator of other, more specific, and often more important variations. Like the Dow-Jones average, the relative sunspot number is used largely because of historical precedent, is subject to rapid daily fluctuations, and is difficult to predict.

The pre-telescopic reports of sunspots are so sparse that it is in practice almost impossible to derive from them any unequivocal information of the past behavior of the sunspot cycle. Nevertheless attempts at doing so have been made for limited periods by attributing naked-eye reports to times of maxima in the sunspot cycle. It may be possible to derive from these early records indications of long-term, secular behavior, as Eddy, and Stephenson and Clark have suggested; however, these interpretations are always subject to questions of intended suppression of sunspot reports (because of their astrological implications) or, in the opposite sense, an unusual interest in them during a given era. Such sociological effects could indeed color the record.[16]

The most recent, and best catalog of naked-eye sunspot reports is that of Stephenson and Clark for China, Japan, and Korea. Where possible, they used original sources, which are most often dynastic records. A total of 141 sightings were documented, between 28 B.C. and A.D. 1604, or an average of about one per decade. Such an average is of little meaning, however, since only six are reported before A.D. 300 and the rest come largely in bunches; 51 of the 141 sightings came, for example, from the particularly active period between A.D. 1077 and 1278, during the Sung and Yuang dynasties in China. According to Stephenson and Clark, these dynasties were particularly noted for diligence and accuracy of astronomical reports.[17]

Another possible index of solar activity that extends nearly as far as the sunspot records can be found in observations of the sun made at times of total eclipse. We now know that the form of the solar corona and the occurrence of prominences at the edge of the sun change systematically with solar activity, or sunspot number. The reliable eclipse record, however, reaches back only to the early eighteenth century, if we seek unambiguous, *physical* descriptions of the eclipsed sun; this period overlaps but cannot extend the sunspot record. Moreover, the most useful eclipse data, from photographic observations, do not begin until late in the nineteenth century, with a dramatic increase in technique and

16 Eddy, "Historical Evidence for the Existence of the Solar Cycle," in White (ed.), *Solar Output*, 51–71; Siguru Kanda, "Ancient Records of Sunspots and Auroras in the Far East and the Variation of the Period of Solar Activity," *Proceedings of the Imperial Academy (Japan)*, IX (1933), 293–296; A Wittmann, "The Sunspot Cycle before the Maunder Minimum," *Astronomy and Astrophysics*, LXVI (1978), 93–97, Eddy, "The Maunder Minimum," *Science*, CXCII (1976), 1189–1202; Clark and Stephenson, "Sunspot Records," 404–409

17 *Ibid.*, 390–399.

in photographic plate sensitivity occurring about 1890. As with historical sunspot data, we can always hope that new discoveries of older descriptions of the eclipsed sun will turn up in historical sources. In this regard it is curious and frustrating that the usable record is so short; Newton and Eddy have raised the question of why there are no descriptions of the structured solar corona before the eighteenth century when it is, to modern observers, so singularly exciting a spectacle.[18]

The next longest direct record of solar behavior is that of the measured diameter of the sun, begun at the Greenwich Observatory in 1750 and continued in several observatories to this day. Although a nearly continuous daily record, it requires considerable care in interpretation; a number of studies have suggested that the shape or the diameter of the sun may vary secularly.

All other direct measurements of solar activity and solar behavior, such as the occurrence of flares, the ultraviolet or x-ray flux, solar magnetic fields, the total solar radiation, the flow of atomic particles from the sun, or the flux of neutrinos from the solar interior have been recorded for at most a few solar cycles in the present century; thus they tell us only of the most recent interval in the life of the sun.

INDIRECT HISTORICAL RECORDS Observations of astronomical phenomena that respond to solar changes can be interpreted as indirect indicators of the past behavior of the sun. In general the most valuable accounts will come, as do sunspot records, from the post-telescopic era, but often useful data exist for far earlier periods.

Measurements of the brightness of the planets and their moons are now made to monitor possible changes in solar radiation, since these objects shine by reflected sunlight; historical planetary data, and especially the photographic plates taken in the last 100 years or so, can serve as an indirect record of solar luminosity. Observations of the zodiacal light—the faint cone of

18 Eddy, "Maunder Minimum," 1197–1199; *idem*, "The Schaeberle 40-Ft. Eclipse Camera of the Lick Observatory," *Journal for the History of Astronomy*, II (1971), 1–22, Robert R. Newton, *Ancient Astronomical Observations and the Acceleration of the Earth and Moon* (Baltimore, 1970), 39; *idem*, *Medieval Chronicles and the Rotation of the Earth* (Baltimore, 1972), 99, 600–601.

light that reaches above the horizon in dark skies at dusk and before the dawn—may possibly tell of changes in solar activity, if correlations found in modern analyses are correct. Accounts of the zodiacal light, called the "false dawn" by Muhammad, are contained in records kept by Arab astronomers in the first millennium A.D. The brightness and color of the moon at times of lunar eclipse have been shown to follow a possible relationship with solar activity. These may be explained by known changes of solar activity on the density of the upper atmosphere of the earth, through which sunlight is refracted to give the eclipsed moon its characteristic copper color. This relationship has been applied to descriptions of the eclipsed moon as early as the seventeenth century. Comet tails change their shape and add characteristic spikes when they are swept by high speed streams of atomic particles directed outward from the sun; thus historical descriptions of comets, one of the more common sky phenomena reported in records from the Orient or Occident, can be interpreted in terms of past conditions of the solar wind. Instrumented records of disturbances in the earth's magnetic field, documented for more than 100 years, provide another source of indirect data because of well-established relationships between solar activity and geomagnetic storms.[19]

By far the most useful of the indirect indices are reports of the aurora borealis and aurora australis, the northern and southern lights. Aurorae are especially valuable because their relationship to solar activity is direct and relatively simple, and because reports of these phenomena are common and extend far back into history. No telescope is required to see an aurora and they have long been noted as objects of awe and superstition.

Displays of the northern or southern lights result when streams of charged atomic particles from the sun interact with the earth's magnetic field, resulting in particle accelerations and collisions with air molecules that then emit light of characteristic green or red or (combined) white color. Since many of these solar particle streams originate in active regions, where sunspots lie, we find a strong correlation between times of high sunspot num-

19 G. W. Lockwood, "Secular Brightness Increases of Titan, Uranus, and Neptune, 1972–1976," *Icarus*, XXXII (1977), 413–430; F. Link, *Eclipse Phenomena* (Berlin, 1969).

bers and times of frequent auroral occurrence. An even better correspondence is found between the occurrence of large solar flares and subsequent auroral displays.[20]

A number of investigators have made useful catalogs of historical auroral reports and these are under steady revision in the light of newer data or more critical assessment. In the catalog of Fritz, one of the earliest and probably the best known, aurora reports are listed from 503 B.C. onward, although the early reports are sparse—fewer than one per year—until the time of the Renaissance. A large part of this early scarcity may be attributed to the difficulty of securing reports from so remote an age; another factor is the strong dependence of aurorae on latitude and the fact that early historical records for the sparsely populated, high latitude auroral regions are rare.[21]

Auroral counts are commonly substituted for sunspot number in historical reconstructions of solar activity. They are, however, an indirect index of limited utility. Comparison of auroral reports with sunspot number in the era of best, modern observations shows a correspondence that is less than perfect and at times almost non-existent. Aurorae are seen, particularly at high latitudes, at times when solar activity and the sunspot number are approaching their lowest levels; moreover, the peak in auroral frequency lags behind the peak of sunspot number by several years. Thus the appearance of frequent aurorae is indicative of high, or else declining, levels of solar activity; a marked paucity of aurorae is almost always the mark of low solar activity; and a moderate number could indicate anything.[22]

Only in the last few years, with the discovery of large open-field regions of magnetic polarity called "coronal holes" on the sun have we come to understand the reason for this less than ideal correlation. Aurora-causing particles from the sun can come from active regions, where sunspots are, but they can also come in high-speed particle streams from coronal holes, where sunspots are generally absent. Coronal-hole aurorae are believed to be more frequent at times of declining solar activity. There is hope of

20 C. Stormer, *The Polar Aurora* (New York, 1955).

21 H. Fritz, *Verzeichniss Beobachter Polarlichter* (Vienna, 1873).

22 Derek J. Schove, "Auroral Numbers since 500 B.C.," *Journal of the British Astronomical Association*, LXXII (1962), 30–35; *idem*, "The Sunspot Cycle, 649 B.C. to A.D. 2,000," *Journal of Geophysical Research*, LX (1955), 127–146.

sorting the two types of aurorae by the terrestrial latitudes at which aurorae are produced. Sunspot-caused aurorae are initiated by higher energy particles that produce their displays at latitudes farther from the poles of the earth. Coronal-hole-produced aurorae are caused by lower energy particles and are seen nearer the magnetic poles of the earth, at high latitudes.[23]

THE MAUNDER MINIMUM In each of the longer records of solar history we can find evidence of possible secular changes in solar behavior that transcend the shorter oscillations of the eleven-year sunspot cycle. In the record of naked-eye sunspot reports and in auroral catalogs these appear as prolonged periods when reports of these events were significantly more, or less frequent than the average. In the curve of annual mean sunspot number (Fig. 2) we can identify similar trends in the range of values reached at peaks of the sunspot cycle. They are not of uniform height but rise and fall under an apparent long-term envelope that may or may not be periodic. After the 1959 maximum, following a run of four cycles of successively higher amplitudes, the annual averaged sunspot number reached an all-time high from which it now seems to be falling. A similar trend occurred nearly 100 years ago, after an obvious long-term minimum in solar activity that persisted between about 1800 and 1820. A more striking and protracted minimum appears in the same figure in the late seventeenth and early eighteenth centuries, if we can believe the reconstructed sunspot numbers for that period of time.

The period of depressed solar activity from 1645 to 1715, known as the Maunder minimum, has been given special scrutiny in studies of solar variability since it represents the clearest case within reach of telescopic records of a significant secular change in the behavior of the sun. One can question, based on our historical reconstructions of solar activity then, whether the eleven-year sunspot cycle continued to operate during the time or whether it was simply so severely depressed as to be hidden in the uncertainty of the available records.[24]

23 Arthur J. Hundhausen, "Streams, Sectors, and Solar Magnetism," in Eddy (ed.), *The New Solar Physics* (Boulder, 1978), 59–134; N. R. Sheeley, Jr., "The Equatorward Extent of Auroral Activity during 1973–1974," *Solar Physics*, LVIII (1978), 405–422.

24 Eddy, "Maunder Minimum," 1189–1202; *idem*, "The Case of the Missing Sunspots," *Scientific American*, CCXXXVI (1977), 80–92.

The period is also used as an historically verified solar anomaly to calibrate the longer, proxy record of solar history found in tree-ring radiocarbon. It is also taken as evidence of a possible solar-climate connection, since the years of the Maunder Sunspot Minimum coincide with a period of extreme cold during the little ice age.[25]

We have recently re-investigated the period by reviewing the historical reports of sunspots and other direct solar observations and by applying as well all available indirect or proxy data. In every case the available facts were found to be consistent with the interpretation expressed by Spörer and Maunder when they called attention to the phenomenon in 1887 and 1890. During this seventy-year span, solar activity, as measured in sunspots, fell to prolonged levels so low as to be wholly unlike the run of solar behavior in subsequent time. Fixing reliable annual sunspot numbers for the period is more difficult, and it may be impossible ever to establish whether in general level they hovered near zero or more nearly ten, or possibly twenty on the scale shown in Figure 2. But it seems clear, from direct historical data and the combined weight of the indirect and proxy data, that for seventy years the behavior of the sun was truly anomalous. Curiously the period coincides almost exactly with the reign of Louis XIV, *le Roi Soleil*.[26]

Telescopes more than adequate to see sunspots were in common use during the time and astronomers were interested in the sun's behavior. In the scientific literature of the day, the unusual absence of spots was often noted, and the discovery of a new one was regarded as a reason for writing a paper. There was a noticeable drop of naked-eye sunspot reports during the time, as recorded in China, Japan, and Korea. The form of the solar corona, described in accounts of the sun at eclipses during the time, does not fit what we now know of its appearance when the

25 *Idem*, "Maunder Minimum," 1189–1202; *idem*, "Climate and the Changing Sun," *Climatic Change*, I (1977), 173–190.

26 F. W. G. Spörer, "Uber die Periodicitat der Sonnenflecken seit dem Jahre 1618 . . ." *Vierteljahrsschrift der Astronomischen Gesellschaft, Leipzig*, XXII (1887), 323–329; E. W. Maunder, "Professor Spörer's Research on Sunspots," *Monthly Notices of the Royal Astronomical Society*, L (1890), 251–252; *idem*, "A Prolonged Sunspot Minimum," *Knowledge*, XVII (1894), 173.

sun is active today. The number of aurorae reported in Europe, America, and the Orient during the time fell to uncommonly low levels.

Some investigators, although they acknowledge a drop in auroral incidence, have questioned the reality of the Maunder minimum as a real solar anomaly since there were aurorae reported during the period. This criticism ignores the fact, mentioned earlier, that auroral incidence is not a perfect proxy indicator of solar activity and that aurorae resulting from recurrent magnetic disturbances related to coronal holes on the sun are now known to occur near the times of the minimum of a normal sunspot cycle. Thus our modern knowledge of solar activity would predict a significant number of aurorae, caused by lower-energy particles in recurrent solar wind streams, during a time of anomalously low solar activity like the Maunder minimum.[27]

Evident in any critical examination of the Maunder minimum is the fact that the prolonged dearth of sunspots between 1645 and 1715 was routinely discussed, and apparently generally accepted in articles and books published from the time of its occurrence until about 1850, when the eleven-year solar cycle was at last established. The general acceptance of a uniform sunspot cycle seems to have erased all belief in an earlier period that did not seem to conform. The episode may say something about the ways of science and scientists, our regard for history when it fails to fit modern experience, and the sudden manner in which the mind of science changes. Before Schwabe's belated discovery of the sunspot cycle, in 1843, there were adamant denials of any periodicity of these features of the Sun. After and particularly following von Humboldt's espousal of Schwabe's finding in 1851, no one seemed to challenge whether the sunspot cycle had always been in operation, and in full force. It was a sudden reversal of opinion much like the one that characterized the acceptance of continental drift in our own time.[28]

27 Link, "Sur l'activite solaire au 17eme siecle," *Astronomy and Astrophysics,* LIV (1977), 857–861; Yu. I. Vitensky, "Comments on the So-Called Maunder Minimum," *Solar Physics,* LVII (1978), 475–478.

28 M. J. Johnson, "On Presenting the Gold Medal of the Society to M. Schwabe," *Memoirs of the Royal Astronomical Society,* XXVI (1858), 196–205; Alexander von Humboldt, *Kosmos: Entwurfeiner physischen weltbeschreinbung* (Stuttgart, 1851), III.

PROXY DATA FROM TREE-RING RADIOCARBON Impersonal diaries of natural events written in annual growth rings of trees offer a potential way of recovering solar history that can reach far beyond the limits set by historical records. Of particular interest to solar history is radiocarbon, an isotope of carbon that enters the leaves of trees as carbon dioxide through photosynthesis and is then deposited in the wood of annual growth. The amount of radiocarbon in the atmosphere, and hence the amount taken each year into trees, varies with time, and one of the factors that causes variation is the sun. Thus in measuring the amount of radiocarbon in the wood of dated growth rings we measure, to some degree, the condition of the sun in the year in which the ring was formed. Since species such as the bristlecone pine live to be several thousand years old, and since continuous tree-ring chronologies now exist for nearly 8,000 years into the past, we have the potential for reading solar history back to the time of the late neolithic and early bronze ages.[29]

It will be an imperfect history, written over with other, unrelated events and blurred in time. Radiocarbon is formed in the upper atmosphere of the earth through the action of high-energy galactic cosmic rays that reach the earth from all directions in space. One of the dominant factors that regulates our receipt of galactic cosmic rays, and hence the production rate of radiocarbon, is the sun; another is the changing strength of the earth's magnetic field. From real-time measurements we know that when the sun is more active, and more spotted, the extended magnetic field of the sun shields the earth from some of the galactic cosmic rays, causing radiocarbon production to fall. When it is less active, as at minima of the sunspot cycle, or during the Maunder sunspot minimum, we receive more cosmic rays and radiocarbon production goes up.[30]

But the radiocarbon is formed at the top of the atmosphere and the trees live far below; between are complex processes that we cannot fully reconstruct in detail. This year's radiocarbon, as gaseous carbon dioxide, makes its way to the trees through slow processes of vertical diffusion and atmospheric circulation; some

29 Eddy, "Historical and Arboreal Evidence for a Changing Sun," in *idem* (ed.), *Solar Physics*, 11–34.

30 Richard E. Lingenfelter, "Production of Carbon 14 by Cosmic-Ray Neutrons," *Reviews of Geophysics*, I (1963), 35–55.

is absorbed in the oceans. In the process it will be mixed and diluted with radiocarbon formed in earlier and later years, so that when it finally enters the trees it will be an amount averaged over the natural variations of several decades. Thus we do not expect to find the signature of the eleven-year sunspot cycle in tree-ring radiocarbon, even though we know that cosmic rays, and radiocarbon production at the top of the atmosphere, show the cycle clearly. We do expect to find evidence of long-term changes in the overall level of solar activity, like the Maunder minimum, which lasted seventy years, or, as a smaller modulation, the protracted solar minimum of the early 1800s.

These effects are clearly present in measurements of radiocarbon taken from tree-rings formed during these periods, giving us confidence in this method of objective solar-history reconstruction (Fig. 3). The Maunder minimum appears as a sharply defined increase of about 2 percent in the amount of radiocarbon, as an integrated effect that lasted 80 to 100 years. It marks the most severe, naturally-caused change in radiocarbon level found in tree-rings that were grown since the time of the telescope. It

Fig. 3 History of Relative Radiocarbon Concentration, from Tree-Ring Analyses.[a]

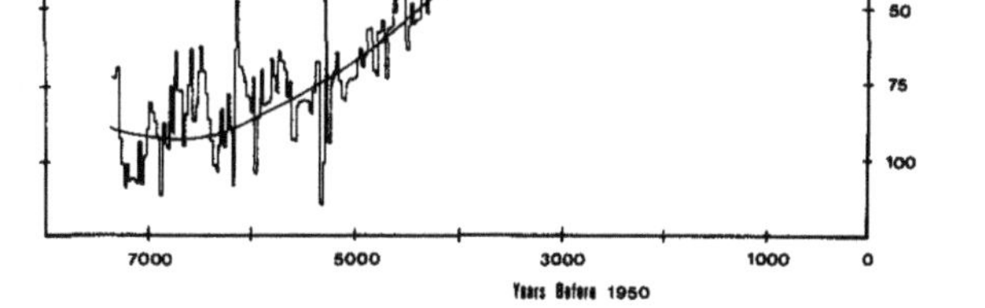

a Plotted are concentrations of ^{14}C relative to ^{12}C, in parts per mil, from data compiled by P. E. Damon, University of Arizona. Increased radiocarbon abundances are plotted downward from the A.D. 1890 norm, which is shown as a dashed line. Solid curve is a sinusoidal fit which matches very closely the recorded change in the strength of the earth's magnetic field. Remaining significant features are of probable solar origin. M = Maunder minimum, S = Spörer minimum, GM = medieval maximum. Eddy, "Maunder Minimum."

confirms our reading of the historical records and offers a yardstick of solar variability with which we can scale the even longer record of tree-ring radiocarbon to identify the nature of other solar effects.

When we do this we find repeated instances like the Maunder minimum in the tree-ring record, all of which come before the time of the telescope and hence beyond the reach of historical records. The most recent of these, before the Maunder minimum, is a similar event, called the Spörer minimum, between about A.D. 1450 and 1540. A less distinct high level of solar activity, which we have called the medieval maximum, is seen in tree-ring radiocarbon records from the twelfth century. It was the last time that the sun was as active as in the present era and, incidentally, a time of warmer climate (known in climatology as the medieval climatic optimum) when the earth was last as warm as now.

The Spörer minimum and medieval maximum of suspected solar activity, scaled from radiocarbon records using the Maunder minimum as a yardstick, show up in records of naked-eye sunspot sightings from the Orient as well as in catalogs of aurora reports.[31]

Eighteen features like the Maunder minimum or the medieval maximum have been tentatively identified in the 7,500-year radiocarbon record from tree-rings. The present radiocarbon record, however, is at best a first approximation, for it comes from compilations by a number of different laboratories and it is under continued improvement as more wood is analyzed and as techniques improve. In time we should have a clearer account that will tell us much more about the real history of the sun.[32]

For now, however, there can be little doubt that the sun changes secularly. And it seems to change in such a way that the present era of solar behavior, in which we have made all our detailed observations of the sun's surface, is not in the long-term an especially representative sample. Times of suppressed solar activity, like the Maunder minimum, seem far from unique in the longer life of the sun. It may be our own age that is unusual.

A POSSIBLE CONNECTION WITH CLIMATE CHANGE These long-term, secular excursions in the level of solar activity, recognized in the frequency of sunspots and aurorae in historical records and

31 Eddy, "Evidence of the Solar Cycle," in White (ed.), *Solar Output*, 57–63.

32 Eddy, "Climate and Changing Sun," 173–190.

farther into the past through the proxy record of tree-ring radiocarbon, occur in a distinctive, irregular pattern. If it shows any periodicity, it is one of about 2,500 years. What is striking, however, is the fit of the pattern to the corresponding record of climate, even though the climate record is at present poorly known (Fig. 4).[33]

The correspondence of the Maunder minimum (1645–1715) with one of the cold extremes of the little ice age has been pointed out by a number of authors as a possible indication of a strong sun-climate relationship, in the sense that when the sun enters a prolonged period of low activity, the earth responds with a cooler interlude. With only one case the connection could be pure co-

Fig. 4 Past Solar Variability (top two curves)
Compared with Climate Reconstructions (bottom).*

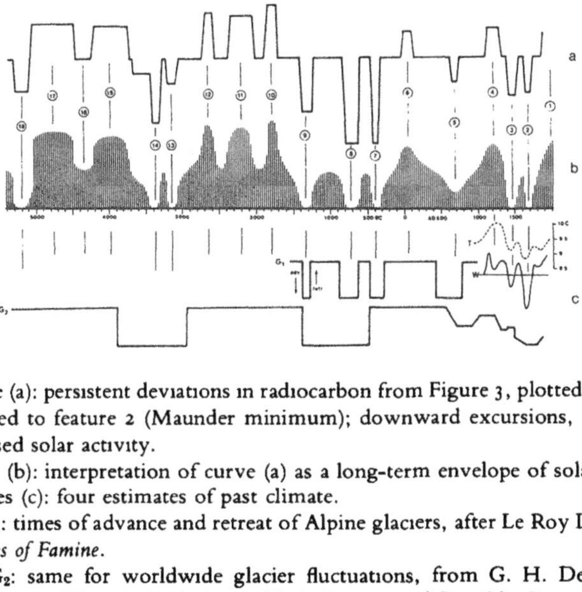

* Top curve (a): persistent deviations in radiocarbon from Figure 3, plotted schematically and normalized to feature 2 (Maunder minimum); downward excursions, as in Figure 3 imply decreased solar activity.
Middle curve (b): interpretation of curve (a) as a long-term envelope of solar activity.
Bottom curves (c): four estimates of past climate.
Step curve G_1: times of advance and retreat of Alpine glaciers, after Le Roy Ladurie, *Times of Feast, Times of Famine*.
Step curve G_2: same for worldwide glacier fluctuations, from G. H. Denton and W. Karlen, "Holocene Climatic Variations—Their Pattern and Possible Cause," *Quarternary Research*, III (1973), 155-202.
Curve T: estimate mean annual temperature in England (scale at right).
Curve W: winter-severity index for Paris-London area from Lamb, "Climatic Fluctuations," in H. Flohn (ed.), *World Survey of Climatology* (New York, 1969), II, 173–249; downward is colder.
From Eddy, "Climate and the Changing Sun."

33 *Ibid.*, 183.

incidence, as is, we can hope, the coincidence of the Maunder minimum with the reign of the Sun King of France.[34]

But, as more of these long-term solar excursions are identified, we can apply a more crucial test. The Spörer minimum recognized in the radiocarbon record and confirmed with weaker evidence in aurorae and naked-eye sunspot reports, coincides with a similar cold dip in climate reconstructions of the little ice age. The medieval maximum of solar activity in the thirteenth century, as established in the same indices, coincides with the medieval climatic optimum, when the climate was last as warm as now. In the sequence of earlier solar excursions derived from the tree-ring radiocarbon record, we find a continued correspondence with climate, as defined by epochs of mid-latitude glacial advance and retreat. The correspondence is of two non-periodic signals, the records of climate and of the envelope of long-term solar variability, and the fit seems almost that of a key in a lock.[35]

We are not accustomed to finding curves of diverse phenomena that fit so well and for this reason we should be cautious in attributing their correspondence to a real sun-earth connection. A real concern is whether the climate itself could be directly modulating the radiocarbon concentration in the lower atmosphere through changes in circulation or in atmospheric and oceanic temperatures. In this case we should expect to find a close correspondence between climatic records and the radiocarbon found in trees. Such a correspondence would leave us, however, with the unexplained correspondence of the Maunder minimum, Spörer minimum, and medieval maximum with climatic deviations, for in each of these cases—the only ones within reach of historical confirmation—we have good evidence of real changes on the sun. Thus the burden of proof of a possible connection between long-term solar changes and climate falls back on the reliability of the historical records, including the weaker ones from pre-telescopic times.

If the connection with climate is real, what is causing it? It cannot be sunspots themselves; moreover the connection with solar activity, if real, seems to ignore the ups and downs of the

34 For example, J. R. Bray, "Solar-climate Relationships in the Post-pleistocene," *Science*, CLXXI (1971), 1242–1243.

35 Eddy, "Climate and the Changing Sun," in *Encyclopaedia Britannica Yearbook of Science and the Future, 1979* (Chicago, 1978), 145–159.

eleven-year sunspot cycle. The simplest explanation, and the one that must be dealt with first, is that we see in each curve—in climate and in the envelope of solar activity—the result of a common, simple cause, which is that of slow, ponderous changes in the total solar radiation, now called the "solar constant." Excursions of but ±1 percent in total solar flux taking place slowly over time scales of centuries could explain the little ice age and other climate features. The same changes in the flow of radiation through the outer layers of the sun could modulate the amplitude of sunspot production, through circulation changes in the atmosphere of the sun, brought about by the action of the solar dynamo. Changes of 1 percent in the solar constant over 100 years would be impossible to detect, even with present instruments. But they might be seen, in retrospect, through their modulation of the peaks achieved in annual sunspot numbers where Spörer and Maunder first found evidence of long-term solar change.

If this explanation is right it could also explain why attempts to correlate climate with the ups and downs of the eleven-year sunspot cycle, or with day-to-day solar activity, have been so frustrating and generally so fruitless, for a slowly-varying solar constant should bear no relationship, other than accidental, to the short-term behavior of solar activity. If this is so we may have been watching the wrong things on the sun for a long, long time.

Thompson Webb III

The Reconstruction of Climatic Sequences from Botanical Data

The temporal sequences of botanical data, ranging from time scales of hundreds of millions down to tens of years, are a rich source of information about past climates. The oldest botanical evidence (spores, leaves, stems, and petrified wood) is fossilized in rocks; more recent evidence (pollen, seeds, needles, wood, and phytoplankton) is preserved in ocean and lake sediments; and the most recent evidence (tree rings and other vegetative material) is still accumulating in living plants. Analysis of these different types of data shows that several vegetative attributes can vary with climate. These include the growth rates and chemical composition of plants, the spatial distribution and abundance of species and genera, and the adaptations of plants to such specific environments as deserts and tropical rain forests.

For each time scale, certain of these vegetative attributes yield clearer climatic records than others. Determining which attributes are best for a given time scale is an important initial task in reconstructing climatic time series from botanical data. On the scale of millions of years, species evolve and become extinct, but the adaptations in plants to certain environments, for example, succulent stems in deserts and entire-margined leaves in tropical forests, have remained relatively unchanged. Fossil floras that show a changing abundance of these adaptive features can therefore indicate climatic changes and yield records of past temperatures. For shorter time scales (1,000 years to 1 million years) when the evolution and extinction of species are less prevalent, the changing abundance and spatial distribution of plant taxa can be a major source of climatic information. On the still shorter time scale of days to centuries, the physiological responses of plants in terms of phenological changes, growth rates, and chemical composition best record the variations in climate.[1]

Thompson Webb III is Associate Professor of Geological Sciences at Brown University.

Grants from the National Science Foundation, Program of Climate Dynamics supported the work on which this article was based. The author thanks Richard H. W. Bradshaw and Sally Howe for their critical reading of an early draft of this article and William S. Stratton and Rosalind M. Mellor for technical assistance.

1 Jack A. Wolfe, "A Paleobotanical Interpretation of Tertiary Climates in the Northern Hemisphere," *American Scientist*, LXVI (1978), 694–703; Harold C. Fritts, *Tree Rings and*

Different types of botanical data are required to monitor the variations in these different vegetative attributes. Each of the data types poses special problems to the paleoclimatologist seeking climatic descriptions in terms of degrees Celsius rather than percentages of tropical leaves. Despite the special problems, a basically similar approach can be followed in calibrating each of the types of botanical data against contemporary climatic data and in finding those aspects of the botanical data that best correlate with certain climatic variables. Ecological evidence and arguments are then sought in order to identify a cause-and-effect relationship between the climatic data and the correlated botanical data. The time series of botanical data that are to be calibrated are also studied in order to establish that their variations reflect the influence of climate. If climate appears to be a causal factor in influencing the past botanical data and if the modern data are correlated with climate, then calibration of the botanical information in climatic terms is straightforward. For quantitative data, in particular, this calibration is amenable to a variety of numerical analyses on calculators or computers.[2]

One means of illustrating exactly how this basic approach to calibrating botanical data works is to describe its application to two particular sets of botanical data, as is done below. Because my own research has focused on calibrating pollen data in climatic terms, the two applications involve climatic interpretation of pollen data that accumulated in the American Midwest over two different time scales, both of relevance to human history. The two examples illustrate the types of climatic changes that have affected the geologically "recent" history of human civilization.

The pollen data in the first example represent the changes in plant abundance during the past 10,000 years and yield 100- to

---

Climate (London, 1976); Thomas M. L. Wigley, Barbara M. Gray, and P. M. Kelly, "Climatic Interpretation of ^{18}O and H in Tree Rings," *Nature*, CCLXXI (1978), 92; H. Arakawa, "Twelve Centuries of Blooming Dates of the Cherry Blossoms at the City of Kyoto and its Own Vicinity," *Geofisica pura e applicata*, XXX (1955), 36–50; Christian Pfister, "The Little Ice Age: Thermal and Wetness Indices for Central Europe," in this issue; Emmanuel Le Roy Ladurie and Micheline Baulant, "Grape Harvests from the Fifteenth through the Nineteenth Centuries," in this issue.

2 Webb and Douglas R. Clark, "Calibrating Micropaleontological Data in Climatic Terms: A Critical Review," *Annals of the New York Academy of Science*, CCLXXXVIII (1977), 93–118; H. John B. Birks, "The Use of Pollen Analysis in the Reconstruction of Past Climates: A Review," paper delivered at the Conference on Climate and History (University of East Anglia, 1979), 5–28.

300-year averages of past climates. These data illustrate some of the environmental changes in the central United States since the time when agriculture arose and cities originated in the Middle East and other areas. Archaeologists and palynologists have long cooperated in gathering and interpreting data on this time scale in order to establish the environmental context of cultural development and expansion.[3]

The pollen data in the second example represent the changes in plant abundance during the past 3,000 years and yield ten- to fifty-year averages of climatic variables. These data indicate climatic changes related to the historically-recorded little ice age of 1430 to 1850 A.D. and show the nature of this climatic event in an area of North America that lacked historical records for most of that time period. Using the climatic information from these and other pollen data along with available instrumental records and climatic data from tree rings and explorer's accounts, meteorologists may soon be able to map the climatic patterns across North America from the twelfth through the nineteenth centuries. These maps will aid the understanding of the extent and causes of the little ice age climatic episode.[4]

POLLEN DATA AND CLIMATIC CALIBRATION Ever since von Post introduced the technique of pollen analysis in 1916, pollen data have been the main source of climatic information on the time scale of 5,000 to 15,000 years. The technique depends upon the steady accumulation of sediments in lakes and bogs to form organically rich deposits that can be collected by hand-driven corers and can be dated by radiocarbon methods. These sediments incorporate a variety of materials including microscopic (20 to 100 μm) pollen grains that have a resistant outer wall. Conifers and flowering plants produce this outer wall to protect the inner

3 Herbert E. Wright, Jr., "The Environmental Setting for Plant Domestication in the Near East," *Science*, CXCIV (1976), 385–389; Johannes Iversen, *The Development of Denmark's Nature Since the Last Glacial* (Copenhagen, 1973)

4 The dates given for the little ice age are from Hubert H. Lamb, "Climatic Fluctuations," in H Flohn (ed.), *General Climatology* (New York, 1969), II, 185. No consensus exists concerning which dates to use for the little ice age, and some scholars even challenge the use of this term to designate a climatic episode. If we use the advance of alpine glaciers beyond their current position as the criterion for defining the little ice age, then evidence for this period exists in most of the alpine regions around the world. See Jean M. Grove, "The Glacial History of the Holocene," *Progress in Physical Geography*, III (1979), 1–54.

sperm-producing cell during sexual reproduction, and wind-pollinated plants produce millions of grains for each one that reaches a receptive stigma. Pollen is both blown and washed into lakes and bogs where the durable walls are well preserved in the accumulating sediments.[5]

In the American Midwest and Northeast, lakes contain 2 to 16 m of sediment that has accumulated in the last 10,000 to 15,000 years since the continental Laurentide ice sheet retreated. Cores of these sediments are subsampled at 5 to 10 cm intervals, and 1 ml samples are processed in the laboratory by a variety of concentrated acids and bases that dissolve away most of the unwanted sediment and leave a residue rich in pollen. Examination of the residue under a microscope at magnifications of 400 to 1000 times permits identification of different types of pollen from the genus to even the species level based on the wall-sculpturing and the shapes of the grains. For each sample, the numbers of grains of each pollen type may be tallied.[6]

Pollen diagrams produced from the tallies show the changing relative abundances of pollen through time at individual coring sites (Fig. 1). The information from several diagrams can then be linked by mapping the different pollen types in similar-age sediments from different lakes (Fig. 2). Patterns in maps of the relative abundances of recently deposited pollen not only resemble the spatial patterns in the plants currently producing the pollen but also are often comparable to the modern-day patterns of climatic variables. In eastern North America, for example, both oak and spruce pollen have north-south distributions that resemble the north-south gradient in temperature (Fig. 2), and herb pollen (excluding ragweed) increases westward with decreasing annual rainfall (Fig. 3).[7]

Maps such as those in Figures 2 and 3 indicate a close correspondence between the distributions of certain pollen types and

5 Lennart von Post (trans. Margaret Bryan Davis, Knut Faegri, and Iversen), "Forest Tree Pollen in South Swedish Peat Bog Deposits," *Pollen et Spores*, IX (1967), 375–401. (English translation of von Post's 1916 lecture.) Edward S. Deevey, Jr., "Living Records of the Ice Age," *Scientific American*, CLXXX (May, 1949), 48–51; R. Bruce Knox, *Pollen and Allergy* (Baltimore, 1979), 1–60.

6 Faegri and Iversen, *Textbook of Pollen Analysis* (New York, 1975).

7 J. Christopher Bernabo and Webb, "Changing Patterns in the Holocene Pollen Record of Northeastern North America: A Mapped Summary," *Quaternary Research*, VIII (1977), 70–71.

Fig. 1 Summary Pollen Diagram from Kirchner Marsh in Minnesota.[a]

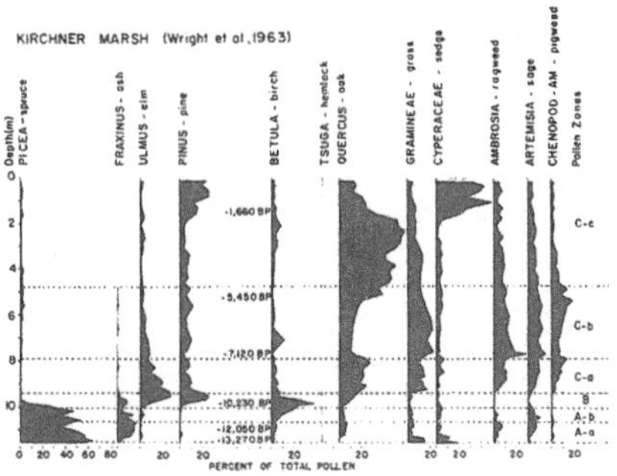

a The radiocarbon dates are given in years before present (B.P.), which is defined to be 1950 A.D. Pollen values are plotted as percentages of the sum of total pollen counted in each sample.
SOURCE: Wright, Winter, and Patten, "Two Pollen Diagrams"; Webb and Clark, "Calibrating Micropaleontological Data," 107.

the distribution of standard climatic variables. When July-mean temperature and the percentages of oak pollen are plotted on a graph (Fig. 4), a strong positive relationship is evident. Existence of such a relationship makes it possible to calibrate the changing relative abundance of oak pollen in terms of numerical changes in temperature. Other pollen types can also be included in this calibration which can then be used to transform past changes in oak and other pollen types into past changes in climatic variables. These calibrations are most accurate for those situations in which the samples of past pollen resemble the samples of modern pollen used in the calibration.

The basic calibration method involves finding a set of weighting factors (B) that rescale the pollen values (P) in terms of the climatic value (C). The model for this procedure can be written as PB = C and represents in symbols the general process used in interpreting pollen data in terms of climate. Such an interpretation results in a transformation of changes scaled in pollen terms (i.e., a 30 percent increase in oak, elm, hickory, and ash pollen) into changes scaled either qualitatively (e.g., warmer and wetter) or quantitatively (e.g., higher by 2°C) in terms of climatic variables. In the situation illustrated in Figures 2 to 4, the quantitative procedure of regression analysis permits estimating the weighting coefficients ($\hat{B}$) directly from the known values of modern pollen

Fig. 2 Maps Illustrating the Modern Distribution of a) Spruce Pollen (as percent of total pollen); b) July Mean Temperature (in degrees Celsius); and c) Oak Pollen (as percent of total pollen).

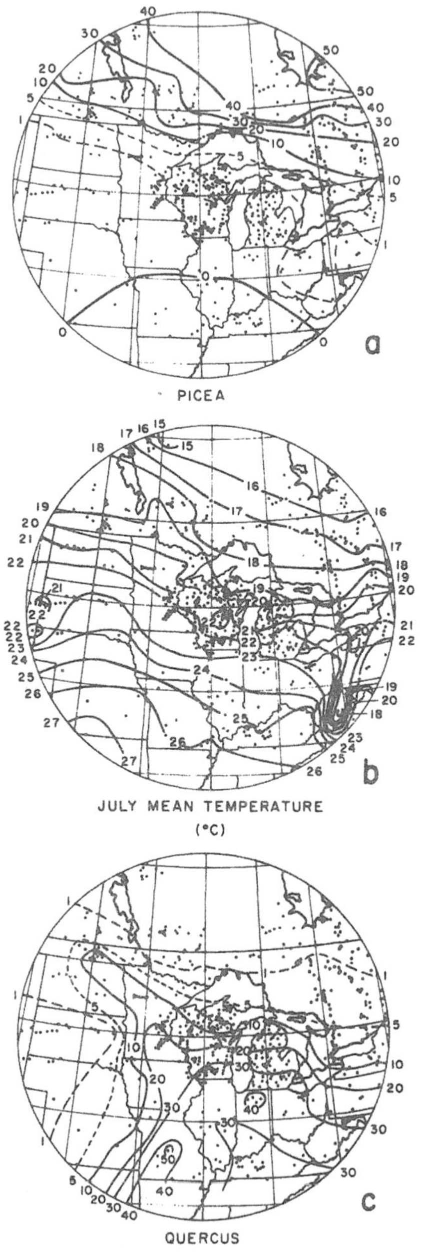

Fig. 3 Maps Illustrating the Modern Distribution of a) Mean Annual Precipitation (in mm for 1941 to 1970 A.D.) and b) Herb Pollen (as percent of total pollen).

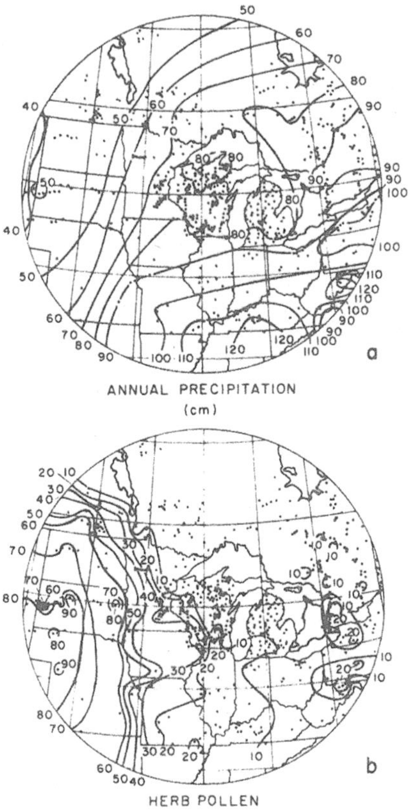

(P_m) and modern climate (C_m), i.e., the B's are calculated from the equation $P_m\hat{B} = C_m$. Estimates of past climate ($\hat{C}_p$) are then calculated by rescaling the values of previously deposited pollen (P_p) by $\hat{B}$, i.e., $P_p\hat{B} = \hat{C}_p$, where $\hat{B}$ and $\hat{C}_p$ are estimates of B and C_p.[8]

The major advantages of this procedure are that it produces quantitative results for which confidence intervals can be calcu-

8 Webb and Reid A. Bryson, "Late- and Postglacial Climatic Change in the Northern Midwest U.S.A.: Quantitative Estimates Derived from Fossil Pollen Spectra by Multivariate Statistical Analysis," *Quaternary Research*, II (1972), 96–98.

Fig. 4 Scatter Diagram Showing the Curvilinear Relationship between the Percentages of Oak Pollen and July-mean Temperature for the 297 Sites in the Region 40 to 52°N and 80 to 95°W.

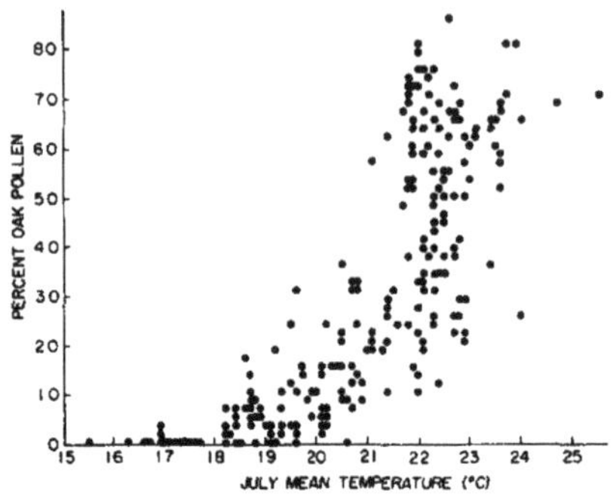

SOURCE: Howe and Webb, "Testing," 154.

lated. The data and procedures are clearly defined and available to all investigators for use, criticism, and refinement.[9]

The assumptions underlying this method of climatic interpretation are also clearly defined and available for criticism. Three of the assumptions are:

1. that no significant changes have occurred to the biological factors in species or genera that affect their competition with other species or genera, i.e., evolution at the species or generic level is insignificant during the time period studied;

2. that modern data provide sufficient information for interpreting past data and, further, that a snapshot of modern spatial patterns provides a sufficient basis for interpreting temporal changes; and

3. that the biological responses (i.e., changes in adaptations, growth rates, or abundances) are and have been related to physical attributes of the biotic environment, in particular to climatic variables.[10]

The main criticisms of the procedure just described have

9 Sally Howe and Webb, "Testing the Statistical Assumptions of Paleoclimatic Calibration Functions," paper delivered at the Conference on Probability and Statistics in Atmospheric Sciences (Las Vegas, 1977).

10 Webb and Clark, "Calibrating Micropaleontological Data," 95–99; John Imbrie and Nilva G. Kipp, "A New Micropaleontological Method for Quantitative Paleoclimatology: Application to a Late Pleistocene Caribbean Core," in Karl Turekian (ed.), *The Late Cenozoic Glacial Ages* (New Haven, 1971), 71–181.

focused on how well the latter two assumptions hold. The concern is that application of the regression coefficients, $\hat{B}$, can yield climatic estimates for pollen changes unrelated to climate or can produce incorrect estimates when spatial variations in modern pollen are not good analogues for past temporal changes. In order to overcome the first of these criticisms, an investigator must establish that climatic changes are the likely cause for the observed changes in pollen.

One way to demonstrate this fact is to produce maps showing the spatial patterns of the temporal changes in the pollen record. Because broad-scale geographical patterns are generally associated with changes in the broad-scale climate, the pollen changes that show similar trends over large areas are probably caused by climatic changes.[11] In contrast, those pollen changes at particular sites that do not occur at nearby sites probably result from one of several nonclimatic factors that influence vegetation. These factors include soil changes, forest fires, human disturbance, local infilling of the site, and invasions by new species.[12]

A second concern with the calibration procedure is that it is empirical and statistical rather than deductive and deterministic. When biologists produce a deterministic model for long-term plant-population changes and base the model on equations derived deductively from well known physical laws that can be assumed to hold constant throughout geological time, then the empirical procedures and their associated shortcomings can be avoided. Because no such deterministic models are available, we have been forced to proceed with empirical methods. In the course of our research, we want to refine the current empirically derived results

11 Bernabo and Webb, "Changing Patterns," 90. Exceptions to this general rule exist and require careful analysis by palynologists. Both the hemlock decline at 4800 B.P. and the recent rise in ragweed pollen have appeared as broad-scale changes on maps of pollen data and seem to have no direct relationship to climate. Pfister, "Climate and Economy in Eighteenth Century Switzerland," *Journal of Interdisciplinary History,* IX (1978), 223–243, has also used the argument that the spatial clustering of data can be attributed to climate rather than to other factors.

12 Birks, "The Use of Pollen Analysis," 10–13; Albert M. Swain, "A History of Fire and Vegetation in Northeastern Minnesota as Recorded in Lake Sediments," *Quaternary Research,* III (1973), 383–396; Iversen, "Retrogressive Development of a Forest Ecosystem Demonstrated by Pollen Diagrams from Fossil Mor," *Oikos,* XII (1969), 35–49; M. B. Davis, "Pleistocene Biogeography of Temperate Deciduous Forests," *Geoscience and Man,* XIII (1976), 13–26; *idem,* "Climatic Interpretation of Pollen in Quaternary Sediments," in Donald Walker and J. C. Guppy (eds.), *Biology and Quaternary Environments* (Canberra, 1978), 35–51.

Fig. 5 Isochrones in Thousands of Years before Present of a) the Northward Decline of Spruce Pollen below 15% from 11,500 to 8000 Years Ago, and b) the Eastward and then Westward Movement of the 30% Isofrequency Contour for Herb Pollen Showing the Movements of the Prairie/Forest Border.

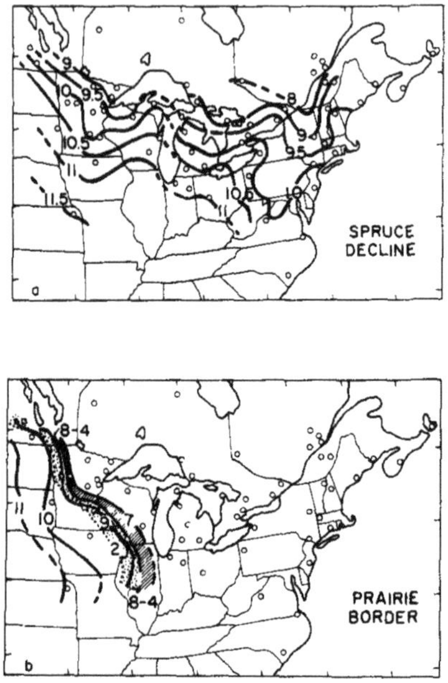

SOURCE: Bernabo and Webb, "Changing Patterns," 84, 89.

and aid the biologists who are developing the deterministic models.

CLIMATIC ESTIMATION: THE FIRST EXAMPLE The 14,000-year pollen record from Kirchner Marsh near Minneapolis, Minnesota, provides excellent data for climatic interpretation. Ample evidence implicates climatic changes as the primary force behind the major changes in pollen values within the diagram (Fig. 1). The decline in spruce pollen 11,000 years ago is a regional event that began about 12,000 years ago in Kansas and moved northward until 8,000 years ago, when spruce values of 15 percent or higher became confined to the current region of the boreal forest (Fig. 5a). This northward movement of the area of spruce dominance reflects the increase in temperature during this time period. The calibration of the record at Kirchner Marsh can therefore provide information on the magnitude of this temperature increase.

The record from Kirchner Marsh also shows a peak in the values of herb pollen (grass, ragweed, sage, and pigweed) from 8,000 to 5,000 years ago (Fig. 1). During this period, prairie extended 50 to 100 km east of Kirchner Marsh (Fig. 5b), which today lies in the region of deciduous forest. Evidence from seeds and diatoms at Kirchner Marsh and at other sites in this area indicate a drop in water levels within the lakes during this period of prairie conditions. Calibration of the pollen record in terms of annual rainfall helps to estimate the magnitude of the climatic change that allowed prairie to replace forest in east-central Minnesota.[13]

Data sets of about 300 sites with both modern pollen and climatic data were used in order to estimate the regression coefficients B that calibrate the pollen data at Kirchner Marsh in terms of temperature and precipitation. Oak pollen is strongly weighted for estimating growing-season temperature (Table 1), whereas the sum of prairie-herb pollen types is highly negatively weighted (Table 1) for estimating precipitation (i.e., increased values of herb pollen decrease the estimated values for rainfall). Time series of precipitation and temperature are then obtained by applying these regression coefficients to the samples at Kirchner Marsh (Fig. 6).

The calibrated record shows a temperature increase of approximately 4°C between 10,000 and 11,000 years ago. This increase is associated with the decline in spruce pollen and the rise in oak and elm pollen (Fig. 1). A decrease of approximately 18 cm in mean annual precipitation was associated with the rise in herb pollen and the prevalence of prairie conditions around the site. These results indicate that between 7,000 and 5,000 years ago the climate in eastern Minnesota was similar to today's climate in western Minnesota and eastern South Dakota. Since 5000 B.P., annual rainfall has increased by approximately 18 cm and tem-

13 Wright, Thomas C. Winter, and H. L. Patten, "Two Pollen Diagrams from Southeastern Minnesota: Problems in Late-Glacial and Postglacial Vegetational History," *Geological Society of America Bulletin,* LXXIV (1963), 1371–1396; Winter and Wright, Jr., "Paleohydrologic Phenomena Recorded in Lake Sediments," *EOS,* LVIII (1977), 188–196. William A Watts and Winter, "Plant Macrofossils from Kirchner Marsh, Minnesota: A Paleoecological Study," *Geological Society of America Bulletin,* LXXVII (1966), 1339–1359.

Table 1 Correlation and Regression Coefficients for Calculating Estimates of July Mean Temperature and Annual Precipitation from Pollen Data.

POLLEN TYPE	CORRELATION WITH JULY-MEAN TEMPERATURE	MULTIPLE REGRESSION COEFFICIENTS	POLLEN TYPE	CORRELATION WITH ANNUAL PRECIPITATION	MULTIPLE REGRESSION COEFFICIENTS
$(\text{Oak})^{1/2}$	.88	0.770	$(\text{Pigweed Family})^{1/2}$	−.71	−0.6733
Spruce	−.71	−0.036	$(\text{Sage})^{1/2}$	−.71	−4.1835
Hickory	.53	0.088	Pigweed Family	−.70	−0.3592
Oak	.81	−0.035	Juniper/Cedar	.20	+0.9683
Birch	−.40	−0.016	$(\text{Hickory})^{1/2}$	.38	+2.0152
			Daisy Family	−.36	−0.4798
			Willow	.22	+0.6244
Constant		18.330	Constant		80.9053
Variance (%)		85	Variance (%)		78
Standard Error		0.8°C	Standard Error		5.4 cm
Area of Samples		40–52°N 80–95°W	Area of Samples		40–47°N 85–105°W
Number of Samples		297	Number of Samples		282

Fig. 6 Estimates of Temperature and Precipitation Derived from the Pollen Data from Kirchner Marsh in Minnesota.

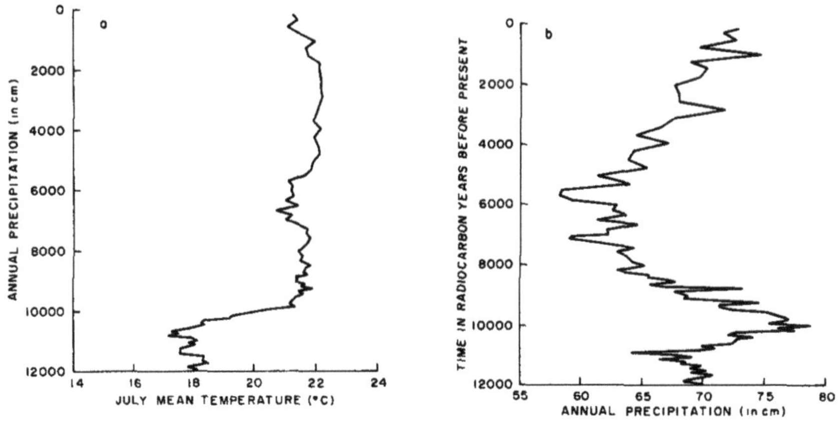

SOURCE: Howe and Webb, "Testing," 156, for temperature estimates.

perature has decreased by 1°C. The climatic changes estimated at Kirchner Marsh thus show that significant climatic changes have occurred over the past 12,000 years.[14]

A comparison of the temperature estimates at Kirchner Marsh with estimates at several other sites in the Midwest shows that the temperature estimates at Kirchner Marsh are consistent with its latitudinal setting. The temperature estimates are higher than estimates at more northern sites and lower than estimates at more southern sites.[15]

The precipitation changes are also consistent with the position of Kirchner Marsh relative to the movement of the prairie/forest border. Precipitation decreased across Minnesota when the prairie/forest border moved eastward and then increased when the prairie/forest border moved back westward (Fig. 5b).

14 Webb and Bryson, "Late- and Postglacial Climatic Change," 106. The final temperature change of 1°C agrees with the indications at many mid-latitude sites for climatic cooling in the last 2,000 years; but, at Kirchner Marsh, the temperature decrease is based on an increase in the values of pine pollen in this time period. This increase may not reflect actual changes in the surrounding vegetation, but rather may be related to the sedimentological changes as Kirchner Marsh changed into a marsh from a shallow lake; Wright, Winter, and Patten, "Two Pollen Diagrams," 1371-1396. When discussing recent changes in climate, one therefore must be cautious in citing the calibrated results from Kirchner Marsh.

15 Webb and Bryson, "Late- and Postglacial Climatic Change," 105.

Precipitation estimates at other sites in Minnesota show the same trend of initial decreasing and later increasing precipitation as is shown at Kirchner Marsh (Fig. 6b). The internal and ecological consistency of these results lends credence to the calibration procedure and the estimates that it produces. Detailed verification, however, using quantitative estimates derived from a set of independent data, is still required.

Even without detailed verification, these time series and maps (Figs. 5 and 6) can be used to show the magnitude and patterns of climatic changes over the past 10,000 years. The calibrated records present a picture of the environmental changes in central North America that were occurring during the period of cultural changes following the development of agriculture in various areas of the globe. The climatic changes elsewhere on the earth differed in detail from those in North America, but the North American record shows just how large and persistent the oscillations in 50 to 100-year climatic averages can be over the scales of millennia and longer.

For the human populations in North America, these long-term climatic changes represented complete alterations of ecosystems at some locations, as illustrated by the transition from forest to prairie at Kirchner Marsh. Today this scale of change would affect both hunting practices, by switching the prey from moose to bison, and agricultural practices, by making it better to grow wheat than corn. Knowledge of the variation in the climatic environment provides a background for understanding certain of the broad-scale human and mammalian changes during the past 12,000 years, such as the role of human hunters in the extinction of large mammals.[16]

CLIMATIC ESTIMATION: THE SECOND EXAMPLE Embedded within the long-term climatic record of the last 10,000 years are several, relatively short neoglacial periods within which alpine glaciers advanced in many areas of the world and the polar pack-ice in the northeast Atlantic Ocean was further south than it is today. The most recent of these periods was the little ice age, which is recorded by meteorological instruments and historical accounts as well as by geological and botanical data. The second example

16 Paul S. Martin, "The Discovery of America," *Science*, CLXXIX (1973), 969–974.

of climatic calibration illustrates the contribution of pollen data to the understanding of this geologically short time period.[17]

Study of the 10,000-year pollen records with their sampling intervals of 200 to 500 years (Fig. 1) has often shown these records to be too coarse to illustrate short-term fluctuations like the little ice age. A different strategy for sampling the pollen data is therefore required if these data are to yield climatic data on such short-term and humanly significant time-scales as 50 to 100 years. This sampling plan has three major requirements: 1) that pollen samples be dated with more precision than the 200- to 1,000-year uncertainties often obtained by radiocarbon dating; 2) that pollen samples be spaced in intervals of twenty to fifty years (i.e., in intervals of 1 to 5 cm in a core) in order to gain the temporal resolution needed to record 80- to 200-year climatic oscillations; and 3) that the sites be in regions in which the pollen data are highly sensitive to climatic changes.[18]

Research following this tripartite strategy is relatively recent and has been spurred on by the discovery and close-interval sampling of lakes with annually laminated (i.e., varved) sediments. In such lakes, dating can be accurate to within five years over periods of centuries or more.[19]

Research concerned with identifying areas with climatically sensitive vegetation has also aided this tripartite strategy. Certain types of vegetation, such as stress-tolerant pine-and-oak-scrub forests on sandy outwash-deposits, can be relatively insensitive to certain climatic fluctuations. Samples are therefore best-suited for recording short-term climatic events if the samples are located near vegetational boundaries or in species-rich communities

17 George H. Denton and W. Karlen, "Holocene Climatic Changes, their Pattern and Possible Cause," *Quaternary Research*, III (1973), 155–205; Grove, "The Glacial History," 41–44; Lamb, "Climatic Variation and Changes in the Wind and Ocean Circulation: The Little Ice Age in the Northeast Atlantic," *Quaternary Research*, XI (1979), 1–20.

18 Bernabo, "Sensing Climatically and Culturally Induced Environmental Changes Using Palynological Data," unpub. Ph.D. diss. (Brown University, 1977), 109–111.

19 Swain, "A History of Fire"; *idem*, "Environmental Changes During the Past 2,000 Years in North-Central Wisconsin: Analysis of Pollen, Charcoal, and Seeds from Varved Lake Sediments," *Quaternary Research*, X (1978), 55–68; Mirjami Tolonen, "Palaeoecology of Annually Laminated Sediments in Lake Ahvenainen, S. Finland: I. Pollen and Charcoal Analyses and Their Relations to Human Impact," *Annales Botanici Fennici*, XV (1978), 177–208; P. G. Appleby, Frank Oldfield, Roy Thompson, P. Huttunen, and K. Tolonen, "^{210}Pb Dating of Annually Laminated Lake Sediments from Finland," *Nature*, CCLXXX (1979), 53–55.

where small climatic changes can shift the competitive advantage from one species to another. Where varved sediment lakes are in areas of climatically sensitive vegetation, their records show well-marked pollen changes on the 200- to 500-year time scale, which is the length of the little ice age.[20]

The difficulty in following the tripartite strategy is that the varved sediment lakes are rare and can lie in regions where the pollen data are relatively insensitive to changes in climate. Under such circumstances, the sampling strategy must be modified in order to unite the information from two sites, one in a climatically sensitive vegetational zone and the other having varved sediments that can aid the dating of the pollen changes at the first site.

Bernabo implemented this sampling strategy in a recent study in northwestern lower Michigan (Fig. 7). He dated a finely sampled, climatically sensitive record at Marion Lake both by close-interval radiocarbon-dating and by stratigraphic correlation with an accurately dated record from a nearby varved sediment lake. The precision of his dating was enhanced because he was able to use the historical date for settlement near the lake. In clearing the land for lumber and agriculture in 1860 A.D., the settlers created conditions favorable to the growth of annual weeds such as ragweed and pigweed. The rise in ragweed pollen from 1 percent to 10 percent in the sediments near the top of the core from Marion Lake indicated the depth in the core that can be dated at 1860 A.D.[21]

The pollen record from Marion Lake shows several changes over the past 3,000 years (Fig. 7), but none are as large or well-

20 For illustration of the differing climatic sensitivity of different forest types, see Linda B. Brubaker, "Postglacial Forest Patterns Associated with Till and Outwash in North Central Upper Michigan," *Quaternary Research,* V (1975), 499–528. Pollen records from varved sediment lakes appear in John H. McAndrews, "Fossil History of Man's Impact on the Canadian Flora: An Example from Southern Ontario," *Canadian Botanical Association Bulletin Supplement,* IX (1976), 1–6; Swain, unpub. data from Clear Pond in New York; *idem,* "Environmental Changes," 55–68.

21 Bernabo, "Sensing Environmental Changes," 107–164. The ragweed rise is a well-marked event in the upper 30 to 120 cm of cores of lake and bog sediments across much of the eastern U.S. and shows that European agricultural practices often polluted the air for hay-fever sufferers long before heavy industry in dense metropolitan areas polluted the air for the rest of the population. See Kent L. Van Zant, Webb, Gilbert M. Peterson, and Richard G. Baker, "Increased *Cannabis/Humulus* Pollen, an Indicator of European Settlement in Iowa," *Palynology,* III (1979), 227–233; Bernabo and Webb, "Changing Patterns," 82–83.

Fig. 7 Pollen Diagram and Estimated Growing Season Temperatures for Marion Lake in Michigan.

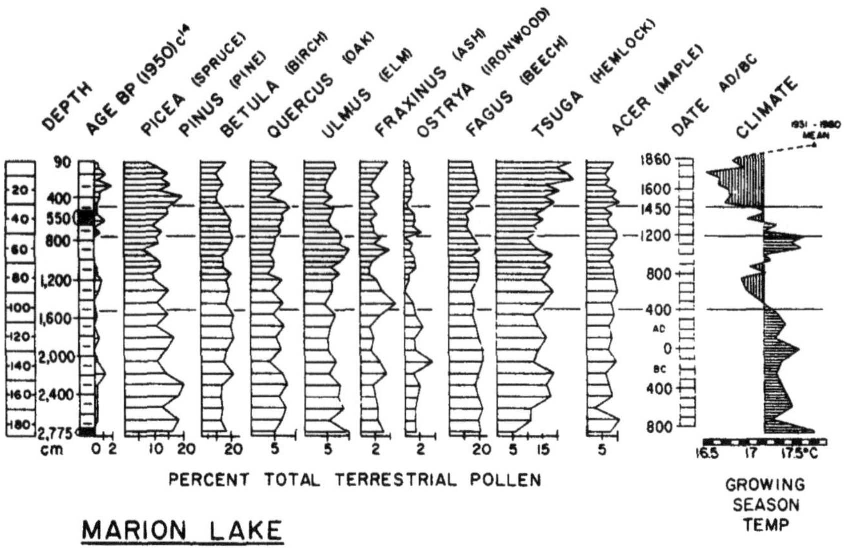

SOURCE: Bernabo, "Sensing Environmental Changes."

defined as those at Kirchner Marsh over 14,000 years (Fig. 1). From 1430 to 1860 A.D., the percentages of hemlock, pine, and spruce pollen were 2 to 10 units above their previous values. These are subtle enough changes to require comparisons with other records before any calibrations are made. The first comparison to be made by Bernabo was with a pollen diagram from Lake 27, which is just 40 km east of Marion Lake and lies in the same vegetational area. The diagram from Lake 27 has a similar rise in hemlock, spruce, and pine pollen in the 500-year period just prior to settlement. The change at Marion Lake is therefore regional for northwestern lower Michigan and not just peculiar to one site.

Comparison of the Marion Lake profile with well-dated and detailed diagrams from more distant sites in Minnesota and from New York shows that all of these diagrams possess a distinctive pollen zone during the time of the little ice age. The nature of the pollen zone varies with the vegetational differences between sites, but the similar timing of the zones across so broad a region suggests that a macroscale climatic change was a likely cause for

Table 2 Correlation and Regression Coefficients for Calculating Estimates of Growing-season Temperature from Pollen Data from Marion Lake in Michigan.

POLLEN TYPE	CORRELATIONS WITH GROWING SEASON TEMPERATURE		MULTIPLE REGRESSION COEFFICIENTS[a]
Pine	−.89		−0.0494
Spruce	−.48		−0.3021
Hemlock	−.63		−0.0845
Oak	.88		+0.0316
Elm	.67		+0.0503
Ash	.70		+0.0379
Constant			64.51
Explained Variance (%)		88%	
Standard Error of Estimate		0.2°C	
Number of Samples		64	

a These regression coefficients yield estimates in degrees Fahrenheit, which are then converted to degrees Celsius.

the changes at each site. Calibrating the record at Marion Lake can therefore indicate what the magnitude of this change was in northwestern lower Michigan.

The calibration requires calculation of regression coefficients that are sensitive to the small magnitude variations in the pollen values at Marion Lake. For this reason, only pollen data from sites in lower Michigan were used in the calculation, and a regression equation showed negative weights for pine, spruce, and hemlock pollen and positive weights for oak, elm, and ash pollen (Table 2).

Application of this equation to the Marion Lake data yielded growing-season temperatures that were approximately 1°C lower than today during the little ice age (Fig. 7). Although there is statistical uncertainty associated with this estimated temperature drop, the 1°C value agrees with the measured temperature departure for 1850 A.D. in northern lower Michigan. The shape of the 2,700-year-old temperature profile, in which two earlier warm episodes appear, parallels the profile at Lake 27, which also shows a temperature range of 1°C. These results suggest that over a large region small changes in climate can have well-marked consequences within the vegetation. In eastern upper Michigan, for instance, Davis has recently shown that beech trees expanded their range 78 km westward during this time period. This change may well reflect the cooler, moister conditions in the upper Midwest during the time of the little ice age. The temperature profile at Marion Lake also parallels other time series from North America and Europe (Fig. 8) and thus adds to the accumulating evidence that the period from 1430 to 1850 A.D. involved a global change in climate.[22]

The two examples presented above illustrate the current state of the art in using regression methods to calibrate pollen data in climatic terms. For both the long and short time scales, the initial results agree with previous qualitative estimates based on ecological reasoning. The general procedure used in both examples involved three steps basic to all methods for interpreting botanical data in climatic terms. These steps include 1) obtaining botanical records from the past and analyzing them to show that they contain responses to climatic variations; 2) studying modern data in order to obtain the weighting factors that can transform the

Fig. 8 Comparison of the Marion Lake Growing-season Temperature Reconstruction with Three Other Paleoclimatic Records.[a]

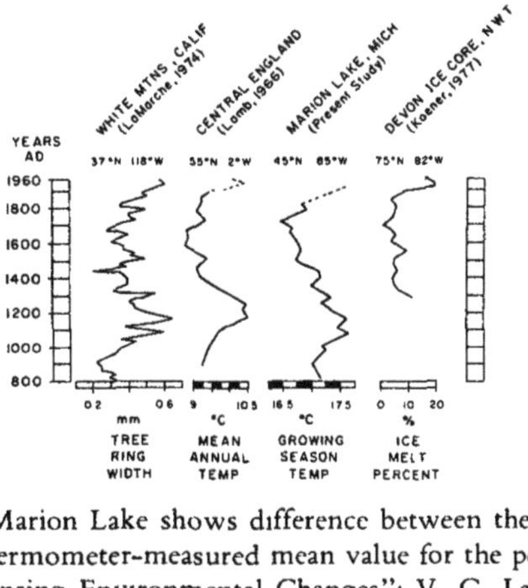

a Dashed section at Marion Lake shows difference between the estimated temperature at 1860 A.D. and the thermometer-measured mean value for the period 1931 to 1960.
SOURCES: Bernabo, "Sensing Environmental Changes"; V. C. LaMarche, Jr., "Paleoclimatic Inferences from Long Tree-Ring Records," *Science*, CLXXXIII (1974), 1043–1048; Lamb, *The Changing Climate* (London, 1966); R. M. Koerner, "Devon Island Ice Cap: Core Stratigraphy and Paleoclimate," *Science*, CXVI (1977), 15–18.

22 Eberhard W. Wahl, "A Comparison of the Climate of the Eastern United States during the 1830s with Current Normals," *Monthly Weather Review*, XCVI (1968), 72–83; M. B. Davis, "Climatic Interpretation," 39.

botanical data into climatic terms; and 3) using these weighting factors on the past data to gain paleoclimatic estimates and then checking these estimates against other pollen records and against paleoclimatic estimates derived from other types of biological data. Numerical techniques such as regression analysis are used in step 2 only for quantitative sets of botanical data.

In contrast to step 2, which involves the regression analysis of geographically distributed modern pollen and climatic data, a different method has been developed and used in Europe for the climatic interpretation of pollen data. The interpretive procedure there uses maps showing the modern range-boundaries of species. A search is made for those species which have boundaries reflecting climatic control. When possible, field observations are also included to demonstrate how climatic factors affect a range of a species. When one or more "indicator" species are identified, then detailed pollen counts are done on fossil material in order to record the presence or absence of the pollen type of this species. From such records, one can infer when a particular climatic variable exceeded or dropped below some limiting value. This approach has shown that temperatures in Denmark 11,000 years ago were 5° to 7°C lower than at present and 6,000 years ago were 2°C higher than today. Birks has critically reviewed the use of this inferential method and provided illustrations of its use.[23]

What makes this alternative approach particularly attractive is that it can be applied concurrently with the regression approach and still provide estimates that are independent of those derived by regression analysis. The independence is obtained because different pollen types are used in the two approaches, and the numerical abundances of the pollen types in regression analysis need not affect the presence/absence of the indicator types and vice versa. The pollen percentages for indicator types tend to be small, whereas the pollen types used in regression analysis have large- to intermediate-sized pollen counts in one or several samples. Interpretations based on the presence of an indicator pollen-type will therefore give an estimate of temperature that is independent

23 Iversen, "*Viscum, Hedera* and *Ilex* as Climatic Indicators. A Contribution to the Study of the Post-glacial Temperature Climate," *Geologiska Föreningens i Stockholm Forhandingar,* LXVI (1944), 463–483; *idem,* "The Late-glacial Flora of Denmark and its Relation to Climate and Soil," *Danmarks Geologiske Undersøgelse,* LXXX (1954), 98; Birks, "The Use of Pollen Analysis," 8–17.

of the estimate based on regression analysis of the major pollen types. In order for this paired interpretative system to work, the European procedure must be implemented in North America. This implementation will require much more detailed pollen analysis and floral studies in North America.

The regression approach is a powerful technique when an adequate data base of contemporary pollen samples is available. Given such a data base, regression coefficients can be derived, and the calibration of the 2,000- to 10,000-year-old pollen data provides reasonable quantitative estimates of past climatic changes. Such results have basically been well received but not without a certain degree of criticism. Most of the criticism concerns whether pollen data accurately reflect past climates. The maps of the broad-scale changes in pollen data provide evidence that climatic change played a major role in effecting changes in the pollen record; but, at specific sites, the worry still exists that biological, soil, and human factors may have a dominating influence on the record. To what degree are the regression equations calibrating changes in the data that are not climatically induced?

One way to deal with this concern is to develop methods that identify and isolate the climatic signal within the pollen record. Finding the climatic signal is not a problem that is unique to the analysis of pollen or other botanical data, however. Just as pollen data record many vegetational changes that are not associated directly with changes in past climates, standard meteorological instruments can record many changes in atmospheric behavior that have nothing to do with changes in the large-scale climate. Highly sensitive barographs with rapidly turning drums can record the effects of opening and closing of the door to the room in which the instrument resides. For meteorological studies, however, the sensitivity and drum-speed of the barograph are set at levels to give accurate records of the passage of high and low pressure systems. A similar practice must be followed in preparing pollen records for climatological analysis. The climatically sensitive part of the record must be separated from the rest of the record.

What is obtained at a particular site from pollen data depends upon many factors including the size and shape of the collecting basin, the rate of sediment accumulation, the time intervals between adjacent samples in a pollen diagram, the time interval during which each sample was deposited, and the number and types of pollen recorded. Choices of what sites to sample and how to sample them influence how easily a climatic signal can be deciphered within a particular pollen record. Much of the information that fascinates a botanist interested in herb pollen from local forests may be of little relevance to the climatologist. Guidelines need to be devised to help paleoclimatologists sort out the forest from the trees. I have recently completed an evaluation of some of the choices that palynologists make when assembling data for vegetational studies. This work must be expanded to include paleoclimatic studies.[24]

When one can assume that the pollen changes reflect climatic changes, the criticism of the calibrations by regression analysis shifts to questioning the accuracy of the calibration of these pollen changes in climatic terms. As long as the assumption that the modern data provide sufficient information for interpreting past data holds, the calibration procedures described earlier work well. For certain modern and fossil conditions, however, the modern data may not be adequate analogues for the fossil data, and inaccurate calibrations may result. This situation arises either because some pollen types are affected by factors today that did not affect these types in the past, or other pollen types were affected in the past by factors that are not critical today.[25]

Human disturbance as recorded in the modern North American pollen record is a major new factor that was not important in the past. Human disturbance includes maintenance of cleared lands for agriculture, selective logging of forests, and control of fires. In eastern North America, the first of these practices led to a shift in the distribution of ragweed pollen from an east-west axis that correlated primarily with precipitation to a north-south axis with a strong temperature dependence. The other two human activities have affected the composition of some eastern North American forests but have not had as severe an effect on the pollen record as the agricultural practices. Because of the human causes of high values of ragweed pollen in modern samples, ragweed

24 Webb, Ruth A. Laseski, Bernabo, "Sensing the Vegetation with Pollen Data: Choosing the Data," *Ecology*, LIX (1978), 1151-1163.

25 Donna C. Admundson and Wright, "Forest Changes in Minnesota at the End of the Pleistocene," *Ecological Monographs*, XLIX (1979), 1-16; Webb and Clark, "Calibrating Micropaleontological Data," 115; M. B. Davis, "Climatic Interpretation," 43-44.

pollen is excluded from the calculations of relative abundances of pollen in modern samples used to obtain the regression coefficients, thus minimizing the bias imposed by recent human activities.[26]

Inaccurate calibrations may also result from those factors that have affected past distributions of pollen, but not today's distribution of pollen. These factors include changes in soil and biological conditions. Certain pollen types have values in 8,000- to 10,000-year-old samples that are higher than any values observed in comparable samples today. The fresh soil on landscapes still receiving glacial meltwaters may be one factor accounting for the higher values in such types as ash and elm. The absence of modern-day competitors such as beech and hemlock may also account for these higher values in the past. Several studies show that the latter two pollen types moved northward and westward more slowly than pine and oak pollen. Climate certainly played some role in the delayed expansion of beech and hemlock; but biological factors, such as seed dispersal and competition from previously established trees, may also have played a role and may make the calibrations for 8,000 to 10,000 years ago less accurate than those for more recent times.[27]

Work is in progress on this problem. At the moment, elm, ash, beech, and hemlock pollen are given little or no weight in the calibrations applied to pollen records covering the last 10,000 years (Table 1). For records of the last 3,000 years, however, these restrictions are not needed, and three of these pollen types are included in the calibration equation for Marion Lake (Table 2).

APPLICATIONS FOR RECENT HUMAN HISTORY Historians may find the work described in this article useful in at least three respects: 1) because the calibrated pollen records document some of the longer term changes in climate that provide a background for

26 The problem of human disturbance is even more severe in Europe and the Middle East, where human disturbance is well recognized in the pollen record from at least 5,000 years ago. The effect of logging on changing of the pollen record was studied by Webb, "A Comparison of Modern and Presettlement Pollen from Southern Michigan," *Review of Palaeobotany and Palynology*, XVI (1973), 137–156.

27 Admundson and Wright, "Forest Changes," 7; M. B. Davis, "Pleistocene Biogeography," 13–26; Bernabo and Webb, "Changing Patterns," 86–88.

understanding recent climatic change; 2) because methodology is described that can be used to derive climatic estimates from written records of wine harvests and barley sales; and 3) because further study of certain botanical records, e.g., pollen from varved sediment lakes, may provide information that historians and climatologists can study in cooperation.

The climatic records of most interest to historians are the short-term, geologically recent records with precise dating. The longer term records, however, should not be neglected, because the changes since Viking times or within the last century are embedded within the longer term changes. The records covering the last 100,000 and 10,000 years tell us that we are living in an ice age during which future advances of the continental ice sheets will occur. These records also tell us that certain areas such as the western Midwest and the Southwest have experienced intermittent droughts for the past 8,000 years.[28]

With the enlargement of the data base derived from varved sediment lakes, an opportunity now exists for a fresh collaboration between historians and paleobotanical-climatologists. These records not only cover the time period of most written documentary evidence, but they also have the temporal resolution and accuracy that most historians desire. For historians concerned with environmental and forest history, the pollen records in varved-sediment records can specify dates of such catastrophes as major forest fires and tree destruction by hurricanes during and before the period of recorded history. Knowledge of these disasters, some climatically induced, can aid in the management of current forests. Similar records of the timing and severity of past droughts have well-acknowledged implications for agriculture and drinking-water reserves.[29]

28 National Research Council: Committee for the Global Atmospheric Research Program, *Understanding Climatic Change: A Program for Action* (Washington, D.C., 1975), 127–195; Bernabo, "Proxy Data: Nature's Records of Past Climates," *Environmental Data Service* (March, 1978), 1–7.

29 Swain, "A History of Fire"; Wright, and M. L. Heinselman, "The Ecological Role of Fire in Natural Conifer Forests of Western and Northern North America—Introduction," *Quaternary Research*, III (1973), 319–328; Jerome Namias, "Severe Drought and Recent History," in this issue; John R. Borchert, "The Dust Bowl in the 1970s," *Annals of the Association of American Geographers*, LXI (1971), 1–22.

Harold C. Fritts, G. Robert Lofgren, and Geoffrey A. Gordon

Past Climate Reconstructed from Tree Rings

Three severely cold winters over the greater part of the United States accompanied by droughts in the first year, and high precipitation in the following two years, suggest to the layman that either our ideas of "normal" climatic conditions are wrong or that climate is changing. Although the evidence for an actual climatic change continues to be vigorously debated, past climate was at least more variable than our present climate. Consequently, there is a renewed interest in assessing to what extent the climate of the past has differed from that of today and what influence it has had upon man.

Modern man has expanded his activities up to and beyond the limitations imposed by his environment. In the Third World, the pressures of increasing populations have forced people still dependent on primitive agricultural practices to occupy regions which have only a marginal capacity for food production. Industrialized societies continue to rely heavily upon technology to compensate for the harsh environment. Hybrid crops and livestock which are developed for high productivity in specific climatic conditions are more highly susceptible to unanticipated climatic conditions than the less productive native species. Global technology and economic strategies have spread through all parts of the world, forming an interdependent network which is increasing in its sensitivity to climatic variations. This sensitivity, coupled with the fact that extreme climatic fluctuations in one region may be linked by atmospheric circulation patterns to extreme climatic fluctuations in other regions, increases the likelihood of disruptions in trade, economics, and politics on a global scale.

Harold C. Fritts is Professor of Dendochronology at the University of Arizona. G. Robert Lofgren is Research Specialist and Geoffrey A. Gordon is Research Associate in the Laboratory of Tree-Ring Research at the University of Arizona.

This research has been in progress since 1970 and presently is supported by National Science Foundation grant number ATM77-19216, Atmospheric Sciences, Climate Dynamics Program. Substantial contributions to this project have been made by Terence J. Blasing and John H. Hunt. These reconstructions of climatic variability are the best estimates available as of January, 1979, and the results are improving as modeling techniques and verification procedures are further developed.

One way to avoid future climate-induced crises is to assess the extremes of past climatic variation and to reserve sufficient food, energy, and other resources to withstand the probable crises that could result from the extreme conditions. However, the length of the meteorological record is inadequate for reliable statistical assessment of decadal or longer period changes that would produce such crises. Few northern hemisphere sea level pressure, precipitation, and temperature records exist before 1850; and prior to the twentieth century, spatial coverage of all types of meteorological data is quite poor.[1]

A promising alternative is to seek evidence for pre-twentieth-century climatic conditions from proxy data, i.e., historical, geological, and biological substitutes for meteorological records, which extend further back in time. Each of these proxies differs in spatial coverage, length, and ability to resolve variations at different time scales. For example, sediment cores obtained from the oceans provide millennia-long climatic records, but the smallest resolvable time scale is several hundred to several thousand years. Pollen data provide shorter sequences with a higher resolution, but are still limited to time scales no smaller than fifty to several hundred years. Layered lake sediments and ice cores of variable length can provide a continuous sequence with a time resolution approximating one year, but they are found in highly restricted localities and are not always dated to the exact year that the layers were formed. Tree-ring chronologies, which can be obtained from all temperate and subarctic forested regions, have a more precise time resolution as they are dated to the exact season and/or year. Historical data on climate also have a similar time resolution, at times to the exact moment that an event occurred, but unlike tree rings they may reflect varying interpretations of past events and often lack continuity over both space and time.[2]

Climatic information derived from well-dated tree-ring records, in combination with the available meteorological data on the recent past and the more extensive historical information, can be used together to reconstruct and assess past variations and extremes in climate.

1 Stephen H. Schneider, *The Genesis Strategy* (New York, 1976).

2 Alan D. Hecht (ed.), Roger Barry, Harold Fritts, John Imbrie, John Kutzbach, J. Murray Mitchell, and Samuel M. Savin, "Paleoclimatic Research: Status and Opportunities," *Quaternary Research*, XII (1979), 6–17.

TREE RINGS AS PROXIES OF CLIMATIC VARIATIONS *Dendroclimatology* is the discipline which uses the annual growth layers of trees called tree rings to study past climate. The size, structure, and chemical composition of these rings can provide climatic information only when the tree growth has been limited by one or more factors of climate throughout a portion of the year.[3]

Generally, one ring is formed each year in temperate climate trees and its width can vary as a function of different limiting factors, including those of climate. The skilled dendroclimatologist can recognize and select trees with potential climatic information from the tree form and from the appearance of the rings. The rings of the selected trees are sampled by extracting a pencil-sized cylinder of wood. A complete sample can include cores from ten to fifty climate-stressed trees growing on similar sites in a local area. The cores are processed in such a manner as to reveal the width and detailed structure of the annual growth layers. Since nonclimatic factors can obscure the effects of climate, the rings are examined, studied carefully, and screened before they are measured.

In particularly unfavorable years, the rings are narrow, and in some trees no ring is formed at all. In other seasons an unusual period of drought during the growing season may cause some trees to produce more than one ring for that year. Therefore, a simple count of the rings from the outside to the center of a living tree is an unreliable method for identifying the year in which each ring was formed. The variability in width and other features among trees must be compared from one year to the next and synchronous patterns in the rings noted as characteristic of unique time periods with particular climatic conditions in certain years.

Using these identifiable ring patterns for certain years, the sizes and appearances of rings are matched with those in other trees to locate the samples and years in which there is a lack of synchroneity, indicating that no ring is visible or that two layers are present in particular places on the sample. After these discrepancies are identified and all ring sequences match correctly, the

3 Fritts, *Tree Rings and Climate* (London, 1976), 434; Valmore C. LaMarche, Jr., "Tree-Ring Evidence of Past Climatic Variability," *Nature*, CCLXXVI (1978), 334–338; *idem*, "Paleoclimatic Inferences from Long Tree-Ring Records," *Science*, CLXXXIII (1974), 1043–1048; Charles W. Stockton and David M. Meko, "A Long-Term History of Drought Occurrence in Western United States as Inferred from Tree Rings," *Weatherwise*, XXVIII (1975), 244–249.

appropriate date is assigned to each ring before the sample is measured and analyzed. This matching procedure, a tedious and time-consuming operation, is called crossdating, and is an essential part of all dendroclimatological work, because it is the only way to identify the exact year in which each ring was formed.

The rings in most trees become narrower with increasing tree age. This systematic change in ring width, which is not climate induced, is eliminated by a computer procedure called standardization. Each measurement of ring width for a particular year is divided by the expected growth for that tree and age on the particular site. The standardized values are then averaged with those of the other sampled trees to obtain mean yearly values of the relative tree growth. A tree-ring chronology is simply the time series of these mean standardized yearly values for a particular site. More than a thousand chronologies are now available for North American conifers and collecting, research, and analyses programs have begun in most temperate regions of the world.[4]

We have selected sixty-five of the best chronologies from western North America for our particular analysis. Each chronology represents a sample of ten or more trees, two cores sampled from each tree from one of the following species: Douglas-fir (*Pseudotsuga menziesii*), bigcone-spruce (*P. macrocarpa*), ponderosa pine (*Pinus ponderosa*), pinyon pine (*P. edulis*), limber pine (*P. flexilis*), bristlecone pine (*P. longaeva*), Jeffrey pine (*P. jeffreyi*), and white fir (*Abies concolor*).[5]

RECONSTRUCTING SPATIAL VARIATIONS IN CLIMATE If a large number of tree-ring chronologies are available for a region and if they represent the effects of a specific climatic variable, the departures from the average tree growth can be mapped, contours drawn, and deductions made as to departures in climate. However, such maps can only be considered unrefined representations of climate. The extent to which climatic factors affect the rate of growth differs among ring-width chronologies so that there is variation in the relative importance to growth of temperature and

4 Fritts, *Tree Rings and Climate*, 261–268; LaMarche, "Tree-Ring Evidence," 334–338.

5 Fritts and David J. Shatz, "Selecting and Characterizing Tree-Ring Chronologies for Dendroclimatic Analysis," *Tree-Ring Bulletin*, XXXV (1975), 31–40; Fritts, Lofgren, and Gordon, "Variations in Climate Since 1602 as Reconstructed from Tree Rings," *Quaternary Research*, XII (1979), 18–46.

precipitation from one month to the next throughout the year. The climate for a given year t may include different factors such as temperature, precipitation, sunshine, and wind (Fig. 1), which vary in their importance to growth and in the sign of their relationship to growth throughout the autumn, winter, spring, and summer months. The growth layer is formed in the spring and summer and may be directly affected by climatic factors at that time. During autumn and winter, the seasons of dormancy in stem growth, both high and low values of climatic factors can have important effects on soil moisture recharge, the making and storage of food in the tree, and the growth of roots.[6]

In addition, there are many lagging effects of climate which appear in the growth ring of the succeeding years $t + 1$ through $t + k$ (Fig. 1). The climate in the preceding year, $t - 1$, controls the formation of leaves and roots along with ring width which is in turn related to the efficiency of the tree in the following years. This produces an autocorrelation in the chronologies; narrow rings tend to be followed by narrow rings, even though the climate of the second year was favorable to growth. All of these interrelationships must be considered when the climatic information for a particular year is extracted from the ring-width sequence.

Dendroclimatologists generally recognize that there has been no simple response to climatic factors. In our work computers have been employed to decode and reorganize the information in a way that directly portrays the variations of individual climatic factors. The ring-width chronologies were compared to meteorological data, and a statistical equation, which provided a type of calibration, was obtained. The calibration equation was then applied to ring-width measurements in the past to reconstruct the corresponding meteorological conditions that must have produced the observed growth (Fig. 2). The actual calibration is a complex, multivariate procedure described in detail elsewhere. Calibration involved correlations and least squares statistical analysis between the meteorological measurements and the tree indices to obtain the appropriate coefficients. There were many

6 Fritts, "Tree-Ring Evidence for Climatic Changes in Western North America," *Monthly Weather Review*, XCIII (1965), 421–443; *idem*, *Tree Rings and Climate*, 377–400, 422–441; *idem*, "Relationships of Ring Widths in Arid-Site Conifers to Variations in Monthly Temperature and Precipitation," *Ecological Monographs*, XLIV (1974), 411–440.

Fig. 1 Model of the Effect of Climate of a Given Year t on Growth in Years t − 1 to t + k.

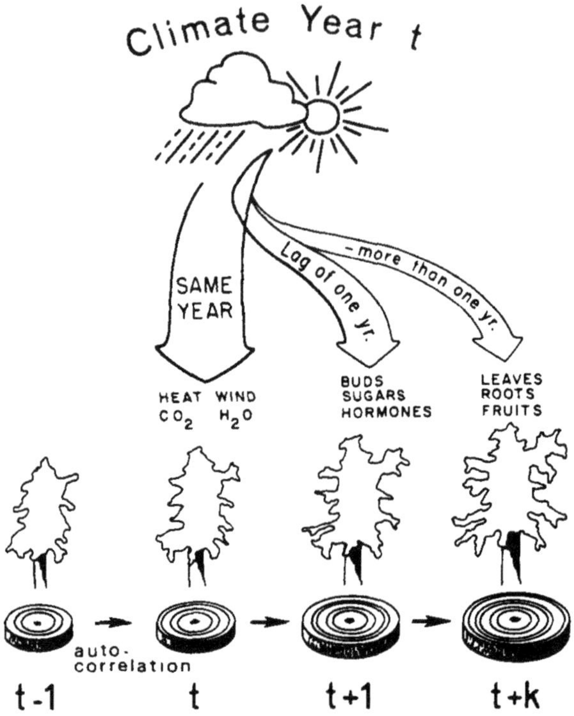

coefficients for a particular meteorological station and chronology. Each coefficient was multiplied with the appropriate tree-ring index for year t − 1 to year t + k and the products summed to obtain an estimate of the meteorological measurement for year t climate. The coefficients were empirically derived and were unique for each meteorological record, season, and chronology used in the calibration. The period 1901–1962, common to both the tree-ring indices and the meteorological measurement, was used for the calibration.[7]

Many tree-ring chronologies were calibrated with many me-

7 Terence J. Blasing, "Time Series and Multivariate Analysis in Paleoclimatology," in H. H. Shugart, Jr. (ed.), *Time Series and Ecological Processes* (Philadelphia, 1978), 213–228; Fritts, *Tree Rings and Climate*, 437–455; Fritts, Lofgren, and Gordon, "Variations in Climate," 18–46.

Fig. 2 Model of the Way the Statistical Calibration Coefficients (b) are Used to Reconstruct Climate in Years t from Tree Growth x in Years t − 1 to t + k.

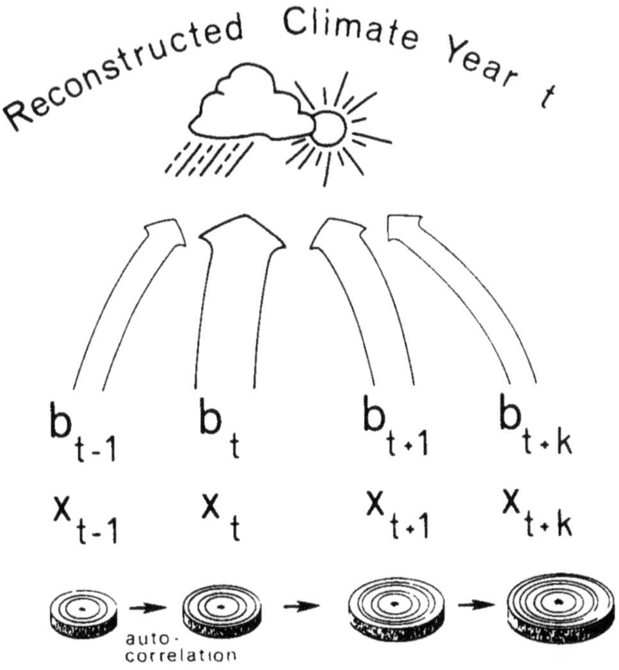

teorological records in one computer analysis by enlarging the number of coefficients and increasing the number of chronologies and climatic stations. Separate calibrations were obtained for different variables, seasons, models, and grids. Because climate is a spatially continuous phenomenon which can be correlated over large areas, climatic stations at distances of 1,000 or more miles from the tree sites, as well as those stations near them, can be calibrated with a ring-width chronology. We calibrated the entire set of sixty-five western tree-ring chronologies with: 1) ninety-six grid points of seasonal sea level pressure at ten-degree longitude and latitude grid squares from latitude 80°W to 100°E and from 20°N to 70°N, except every 20° of longitude at 60° and 70° latitude north; 2) seventy-seven seasonal temperature records from stations in the United States and southwestern Canada; and 3) ninety-six seasonal precipitation records over the same area as temperature. The seasons we used were: winter: December–February; spring: March–June; summer: July–August; and autumn: September–November.

Each calibration produced a large array of coefficients which were assembled as a transfer function. The coefficients in the transfer function were multiplied by the sixty-five tree-ring chronologies at the appropriate time lags to obtain seasonal estimates for the climatic variable at each climatic station called the reconstructions. The transfer function was applied backward in time on earlier and earlier tree rings to obtain the corresponding yearly variations in climate for those years prior to the period of calibration. The reconstructions for each season and year could then be accessed by computer and mapped or plotted in a variety of ways to portray the patterns in past climate.

An important statistic used to evaluate the success of a calibration was the percent of variance calibrated, that is, the percentage of reconstructed variations that resemble those in the actual meteorological data during the calibration period. The models used in this article were selected largely on the basis of this statistic. The technical details and development of the different models will be treated elsewhere.[8]

VERIFICATION WITH INDEPENDENT DATA The calibration coefficients are most reliable if they are calculated using many years of meteorological and tree-ring data. However, it is easy to overmanipulate data so that the calibration appears successful, but does not accurately reconstruct climate for years outside the interval used for calibration. It is necessary to reserve some independent data, i.e., data not used in calibration, for verification (or validation) of the reconstructions.

Verification involves testing the reconstructions against other information on climate for years outside the interval used for calibration. This information includes independent meteorological measurements, related proxy information, and various types of historical evidence. Each type of information has its strengths and weaknesses, and all three can be used to verify different features of the reconstructions.

8 Fritts, Lofgren, and Gordon, "Tree-Ring Reconstructions of Past Climatic Variations," in typescript.

Table 1 Summary of Verification Test Results for a Spring Temperature Model

TEST	PERCENT PASSED DURING: INDEPENDENT PERIOD	CALIBRATION PERIOD
Correlation coefficient	30	100
Correlation on first differences	49	100
Sign agreement	9	99
Sign agreement on first differences	21	97
Product means	19	99
Percent of total tests passed	26	99
Reduction of error greater than zero	38	100
Chi-square for pooled data	17	95

Verification with independent meteorological measurements is the most objective and precise method of validation, and for that reason it is of greatest scientific value to dendroclimatology. In our work the independent meteorological measurements were compared directly with the reconstructions and the similarity was measured with five objective statistics that could be tested for significance (Table 1). The results of these tests were expressed as the number or percentage of stations passing each test. The percentage of stations passing the tests over the 5 percent expected by chance supports the contention that the models did contain valid climatic information. A statistic, called the reduction of error, which was analogous but not identical to the percent variance calibrated, was also calculated for each station having independent data. The independent data from all stations and all years were pooled and a contingency table of nine equally probable classes was constructed and tested using the chi-square statistic.[9]

Table 1 includes the results of one verification using a spring temperature model which calibrated 45 percent of the variance. Tests were made on both the independent and calibration periods. There were fifty-three climatic stations tested, each with seven or

9 Edward N. Lorenz, "An Experiment in Nonlinear Statistical Weather Forecasting," *Monthly Weather Review*, CV (1977), 590–602; Fritts, *Tree Rings and Climate*, 332; Hans A. Panofsky and Glenn W. Brier, *Some Applications of Statistics to Meteorology* (University Park, Pa., 1965), 53–58.

Fig. 3 Reconstructed and Measured (real) Spring Temperature Departures for Santa Fe, New Mexico, for the Independent and Dependent Periods.[a]

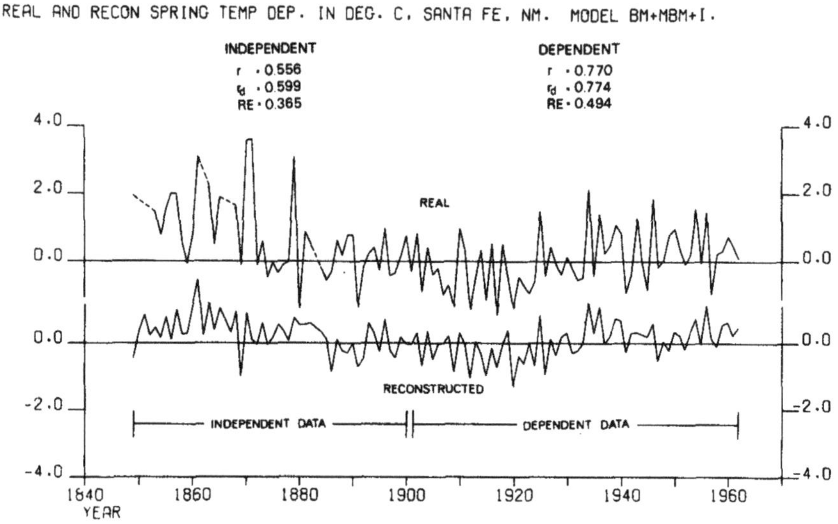

a Verification statistics shown for these two periods are: r—correlation coefficients; r_d—correlation of the first differences; RE—reduction of error.

more years of independent data and an average length of twenty-five years. In this example, twenty-six percent of all the verification tests on independent data were passed. Almost all were significant for the calibration period. The statistics measure different attributes of the reconstructions and an examination of the individual test results can be used to assess particular strengths and weaknesses of the reconstructions.

Figure 3 is a plot of both the reconstructed and measured temperature departure for Santa Fe, New Mexico, a station which passed all five verification tests on forty-four years of independent data. The magnitudes of the reconstructions are smaller than the actual climatic measurements, but the general trends are similar to the measurements. In this example, the long-term trends apparent in the observational data are especially well reconstructed.

ANALYSIS OF RECONSTRUCTED CLIMATIC VARIATIONS We have used reconstructions of past climate in a number of ways. In each example we utilized the best available data—all are derived by applying the transfer function to the same sixty-five tree-ring chronology set. Where possible, we have selected examples for which written historical information was available.

Climatic Variations in Specific Seasons

The calibration for different seasons and variables were made separately and estimates for each climatic station and grid were obtained for every year from 1602 to 1962.

Figures 4 and 5 show the reconstructions for spring and summer of 1849 for which there were military records and diary accounts of migrants traveling to Oregon and the gold rush in California. The military precipitation records showed that the spring of 1849 was relatively wet in western Missouri and Arkansas. The temperature records suggested that spring warming occurred earlier along the Santa Fe Trail than along the Oregon Trail, but travelers on neither trail experienced extremely warm late-spring temperatures. The diary accounts were more subjective but suggested that the migrants were buffeted by frequent storms during late spring and summer.[10]

Fig. 4 Reconstructed Sea Level Pressure, Temperature, and Precipitation for Spring of 1849 Expressed as Departures from the Twentieth Century Means.[a]

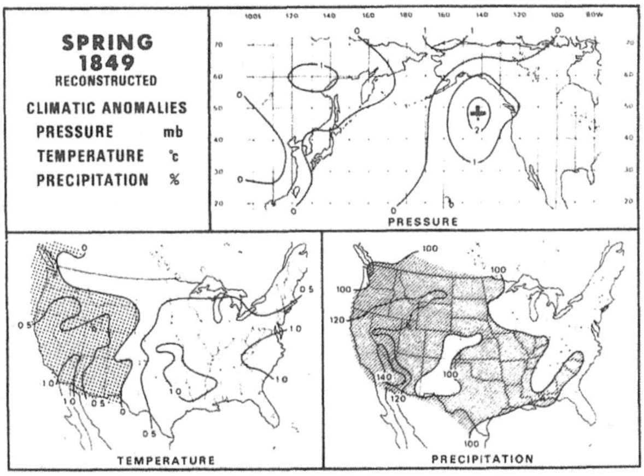

a Mean for 1899-1970 in the case of pressure and for 1901-1970 in the cases of temperature and precipitation.

10 Merlin P. Lawson, *The Climate of the Great American Desert: Reconstruction of the Climate of Western Interior United States, 1800–1850* (Lincoln, Neb., 1974).

Fig. 5 Reconstructed Sea Level Pressure, Temperature, and Precipitation for Summer of 1849 Expressed as Departures from the Twentieth Century Means.[a]

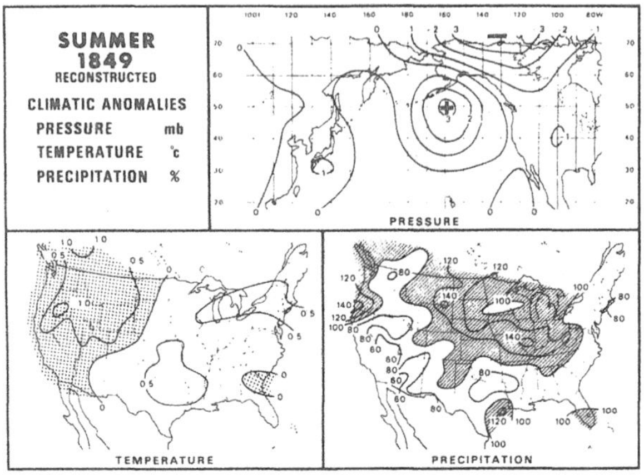

a Same as Figure 4.

Our surface pressure map for spring, 1849 (Fig. 4), shows a weak positive departure pattern which was interpreted as a more intense subtropical high displaced a few degrees north of its normal position in the North Pacific, and a negative departure over the Alaskan and Canadian Arctic, indicating more storm activity there. This negative departure became stronger in summer and extended eastward to Hudson Bay (Fig. 5). The subtropical high was stronger and shifted northwest of its normal position. A steeper pressure gradient was reconstructed over Alaska and western Canada which was interpreted as creating a stronger than normal flow of cold Arctic air into the Canadian Arctic region. The presence of cooler temperatures there was independently confirmed by the narrow rings in many trees growing in the subarctic region. The pressure departures were also inferred to have enhanced a northerly flow of cold air into the Far West and a southerly flow of warm, moist air from the Gulf of Mexico into the prairie states, where the migrants were traveling.

The precipitation reconstructions over the United States for spring show the West to have been wetter than average and the East to have been drier. For summer, the precipitation for the central prairie states was reconstructed to have been as much as 40 percent above the 1901–1970 averages. This reconstruction also agreed with early fort measurements of precipitation which indicated a wet spring and summer. The temperature departures for spring and summer were similar. A cooler West than East was evident, and the positive departures in the lower Mississippi Valley were not as marked in summer.

In this example we have used spatially discontinuous historical climate data and written records to evaluate our reconstructions of pressure, temperature, and precipitation derived from tree-ring data. However, the dendroclimatic reconstructions can be used to interpret the historical data in terms of past climatic conditions and storm systems passing through the country. Both types of data are easier to interpret together than by themselves, and provide the most detailed picture of past climatic patterns.

Regionally Averaged Climatic Variations

When the reconstructions were averaged over space or time for the calibration period, the percent variance calibrated in climate usually increased because the trees themselves are integrators of climate. This increase became apparent when average percent variance calibrated for each year at individual climatic stations or grid points was compared to the same data after averaging the reconstructed and measured climatic data over space or time.

For example, there were seasonal precipitation reconstructions for nine stations in the Columbia Basin which individually had a percent variance calibrated near 41 percent. When these nine reconstructions were averaged together and compared to the average of the nine sets of observed climatic data (Fig. 6A), 54 percent of the variations in the averaged sets were calibrated. The averaging increased the percentage of associated variation by about 13 percent. The averaged data were then smoothed into approximate decadal values by using a low-pass filter, which is similar to a moving average (Fig. 6B). This resulted in 74 percent of the variance being calibrated, an additional increase in the percent association of 20 percent. In this example there was a total change of nearly 33 percent in the relative amount of agreement after averaging over space and averaging over time by filtering.

Fig. 6 Averaged Reconstructions of Winter Precipitation over Nine Meteorological Stations near the Columbia Basin.[a]

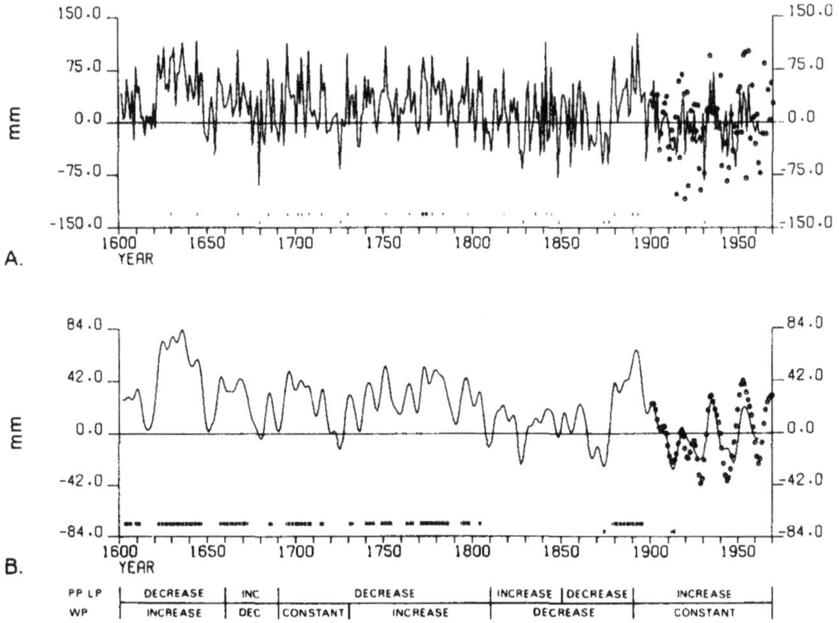

a Data are expressed as departures from the 1901-1970 observational record. Plot A is unfiltered, and Plot B has been treated with a low-pass filter. The observational record from 1901-1970 is plotted as dots on the right. Year designations are for January of each winter season. Intervals of change in the percentage of pollen content of *Pinus ponderosa* (PP), *P. contorta* (LP), and *P. monticola* (WP) observed in the varved sediments from Gillette Lake, Washington, are indicated below the filtered time series.

Another type of verification which we used were the changes of pollen content observed in varved (annually layered) sediments from Gillette Lake, Washington, also shown in Figure 6. The changes in *Pinus ponderosa* (PP) and *P. contorta* (LP), which grow in dry habitats, were diagramed above those for *P. monticola* (WP), which grows better in wet habitats. The decreases or increases in these two contrasting pollen types were in agreement with the long-term reconstructions of lower-than-average and higher-than-average precipitation. There were two intervals at the beginning of the eighteenth century and in the twentieth century, when the percentage of *P. monticola* pollen did not change at all. These intervals were reconstructed to have had near-average precipitation when there would have been less likelihood for a change in vegetation and pollen content. Although the time resolution of the pollen data was markedly less than that of the tree-ring data, the information in both data sets confirmed the presence of the same long-term moisture variations.[11]

The average reconstructed winter precipitation for nine stations in the valleys of California is shown in Figure 7. The regionally averaged meteorological data were plotted to the right. Not all features were coincident in California and the Columbia Basin. For example, a long-term downward trend in precipitation from 1602 through the 1880s was apparent only for the Columbia Basin. The latter half of the eighteenth century was wet in the Columbia Basin and dry in California.

Lynch developed indices of annual precipitation for southern California using mission records from 1770 to 1832 which provided a type of verification. From 1833 to 1850 his data were

Fig. 7 Average Reconstructions of Winter Precipitation over Nine Meteorological Stations in California.[a]

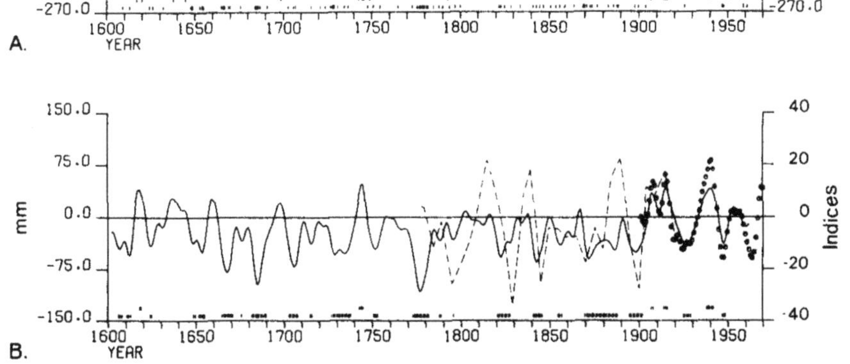

a Dashed lines indicate filtered annual rainfall indices derived by Lynch for Southern California plotted at five-year intervals for 1775-1925. Year designations are for January of each winter season. Lynch, "Rainfall and Stream Run-off."

11 Pollen data are from Albert M. Swain (Univ. of Wisconsin), in litt.

based on historical references to weather conditions, floods, droughts, and crops, but after that date actual precipitation measurements were also used in deriving his indices of precipitation. Filtered values of Lynch's data are shown in Figure 7B as dashed lines joining the points plotted at five year intervals. We calculated our five verification statistics using Lynch's yearly indices for southern California and our winter reconstructions for the entire state of California. All five verification statistics were significant at the 95 percent level for 1770–1900, but only one test was significant when the calculations included only 1770–1830. Comparisons of the results for filtered and unfiltered sets indicated that the 1770–1830 indices, which were based on mission crop records, agreed with the short time-scale variations of our reconstructions but disagreed with the long time-scale values. The indices for 1831–1850 agreed with long time-scale variations better than with the short time-scale variations in our reconstructions. After 1850 there was general agreement at all time scales.[12]

Because there were no obvious differences in the tree-ring sets for the early time intervals, we concluded that the long time-scale changes in the mission records were influenced by social or political factors that were uncorrelated with climatic variation and were therefore poor indicators of past climatic conditions. However, the data used by Lynch for 1831–1850 appear to have followed the long time-scale climatic changes more accurately and better portray the actual climatic variations. The indices improved after 1850 with the availability of some precipitation measurements. The three large peaks in Lynch's data in the nineteenth century were not reproduced well by these particular dendroclimatic reconstructions, indicating that this particular model and calibration was underestimating precipitation during the wet periods. Recent calibrations which were not available in time for this article follow the wet periods more reliably.

Temporally Averaged Climatic Variations

In order to utilize the apparent increased reliability of the reconstructions averaged over time, maps can be made for decades, half centuries, and longer time intervals. Figure 8 is one

12 H. B. Lynch, *Rainfall and Stream Run-Off in Southern California Since 1769* (Los Angeles, 1931), 6–7.

Fig. 8 Mean Deviations of Tree Growth and Mean Reconstructed Deviations in Temperature and Precipitation for the Winters and Springs of 1861-1870.[a]

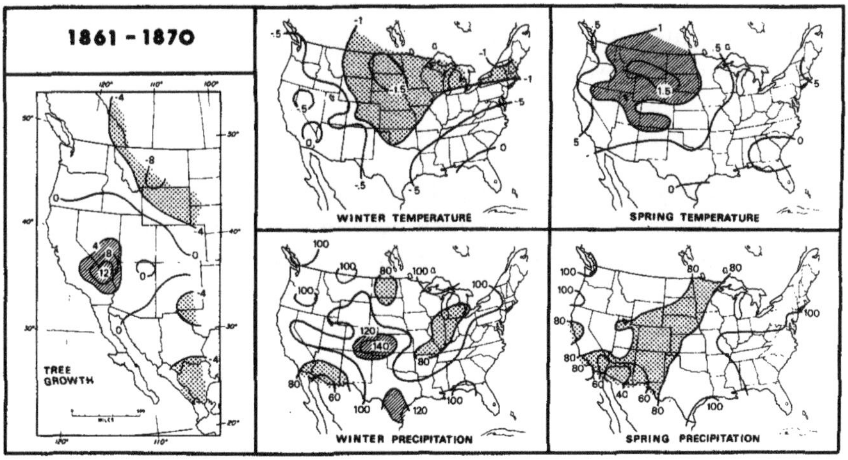

a Tree growth is expressed as normalized values multiplied by 10 using the 1601-1963 means and standard deviations. The climatic reconstructions are expressed as departures in °C or as percentages from the mean period 1901-1970.

example of the averaged reconstructions for which we have some reliable historical evidence in the decade from 1861 to 1870. Below-normal winter temperatures were reconstructed throughout most of the United States and southwestern Canada for this decade, with wetter than normal conditions centered in Colorado and Kansas westward to northern California, and drier than normal conditions in the Great Lakes, the Mississippi and Ohio River valleys, and the extreme Southwest. Spring was reconstructed to have been warmer and drier than the twentieth century, especially in the West.

An examination of a few available historical references yielded no account of unusual temperature for the entire decade. However, several accounts of heavy precipitation were found. The Great Salt Lake was reported to have risen ten feet during this decade. There was one reference to a heavy snowstorm in San Francisco during the winter of 1869 and to heavy rains in Utah in December 1867, causing very high river discharge. During the winter of 1862–63, there was a reference to a drought in Los Angeles, and two references to lower than average precipitation. Although there are too few data to validate these reconstructions fully, the historical references do illustrate the potential for comparison and show how the tree-ring reconstructions can help to reconcile reports of high precipitation in one area (Great Basin and central California) and drought in a neighboring area (the extreme Southwest).[13]

Departures from the Twentieth-Century Normal Figures

Climatologists have warned that the interval from 1930 to 1960 was unusually mild for large portions of the northern hemisphere. The 361-year-long tree-ring reconstructions provide us with a means of measuring how unusual the twentieth century climate is for an area with little or no early meteorological or historical data and when, where, and in what season climatic statistics have varied from seventeenth through twentieth century values.[14]

In order to compare the statistics of the past to those of the present, we chose 1901–1962 to be our representation of the normal period and averaged the reconstructions within eleven separate regions over the United States and southwestern Canada.[15]

The departures for winter temperatures (Fig. 9) show that past winters were reconstructed by this model to have been colder for nine of the eleven regions. Region 3, the Intermountain Basins, was the only region to have had a positive departure. Region 7, the Northern Prairies, had the greatest departure of −3.1°C. In summer the temperature departures were reconstructed to be

13 Ernst Antevs (ed. J. K Wright), *Rainfall and Tree Growth in the Great Basin* (Baltimore, 1938), Plate I, 50; David M. Ludlum, "A Century of American Weather. Decade 1881–1890," *Weatherwise*, XXIII (1970), 131–135; California Historical Society, "The Memoirs of Lemuel Clarke McKeeby," *California Historical Society Quarterly*, III (1924), 169; Lynch, "Southern California," 31.

14 Reid A. Bryson and F. Kenneth Hare, *World Survey of Climatology: 11, Climates of North America* (New York, 1974), 38–47; Bryson, "A Perspective on Climate Change," *Science*, CLXXXIV (1974), 753–760; Hubert H. Lamb, "Understanding Climatic Change and Its Relevance to the World Food Problem," paper presented at the Sixth G. E. Blackman Lecture (Oxford, 1976).

15 We adjusted each reconstruction so that the means and standard deviations of the reconstructions were equal to those of the actual data for the normal period by adding the difference between the means and multiplying by the ratio of the standard deviation of the actual data to the standard deviation of the reconstructions for the normal period. We then calculated the means and standard deviations for 1602–1900 and expressed them as departures or percentages of the 1901–1962 mean values.

Fig. 9 Mean Winter and Summer Temperature Departures for 1602-1900 Expressed in °C from the Mean Period 1901-1962.

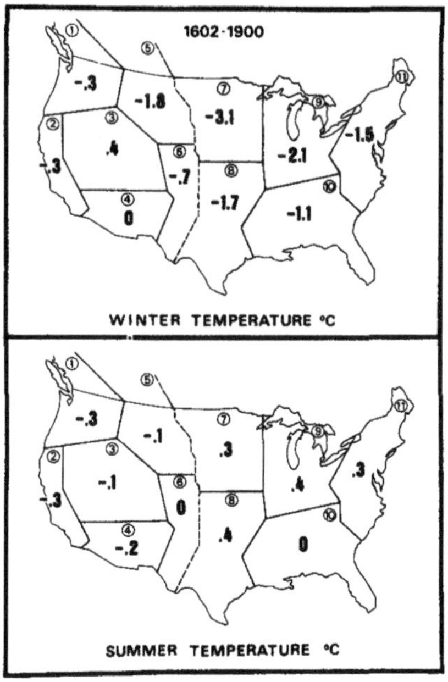

smaller, with slightly warmer conditions in the East and Central Plains and cooler in the western areas.

Precipitation estimates (Fig. 10) totaled over all four seasons were below those of the twentieth century for most western states. The deficit was 15 percent of the 1901–1962 mean for Region 2, California. In the Pacific Northwest and in the eastern third of the country, precipitation was estimated to have been 2 percent to 8 percent higher than the 1901–1962 figures.

Although the values on these maps have changed somewhat as new and better models were developed, the patterns of departures between the statistics of the past and those of the present were reconstructed to be similar. The world-average cooler temperatures of the nineteenth century reported by Willett and extended to earlier centuries by others appear in our reconstructions to have been largely a winter phenomenon restricted to the central and eastern United States. It was not uniform over the entire country and for every season. The three-century trend was in the opposite direction for Region 3. Summer temperatures for the

Fig. 10 Total Annual Precipitation for 1602-1900 Expressed as Percentage Departures from the 1901-1962 Mean.

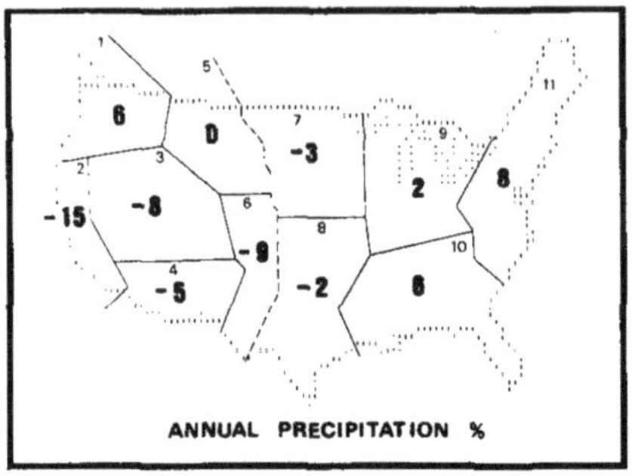

past were reconstructed to have been warmer, not cooler, in the eastern half of the country. Our reconstructions also suggest that the twentieth-century climate has been wetter than in the past. If we were to use the statistics of our reconstructions based upon the seventeenth through nineteenth centuries to project future probabilities, California, which already has had water shortages, on the average would be most likely to have 15 percent less winter moisture than it has had from 1901 to 1962. Such analyses can help us spot regions and identify seasons for which the twentieth-century records are the poorest representations of the longer-term probabilities and for which we may need to make allowances in our long-range planning.[16]

Spatial variations in climate can be reconstructed from tree-ring width data using computer calibrations if there are sufficient numbers and spatial coverage of replicated, well-dated, and standardized tree-ring chronologies. The tree-growth responses to particular climatic variables and seasons vary among the chronologies. These differences can be utilized by calibrating large arrays of chronologies with large arrays of a particular climatic variable in a multivariate analysis. Each variable, season, and array requires a separate calibration. Reconstructions must be compared to available independent meteorological measurements, related proxy records, or historical data to demonstrate their validity.

16 H. C. Willett, "Temperature Trends of the Past Century," in *Centenary Proceedings of the Royal Meteorological Society* (London, 1950), 195–206.

There is a variety of information in the dendroclimatic reconstructions, but large regional variations with high and low centers over the nation vary in importance and sign over the 1602–1962 time period. The reliability of these reconstructions is greater for regional and decade-long averages than for individual station reconstruction of the seasonal climatic variability.

Not all of the United States was reconstructed to have been cooler in the past than in the twentieth century. The greatest departures from the twentieth-century figures indicate markedly colder conditions in the north-central states in winter and drier conditions throughout large parts of southwestern and central North America. In the eastern portions of the country the precipitation was reconstructed on the average to be greater than the twentieth-century figures.

More important for historians, the dendroclimatic reconstructions can provide a well-dated, objective, and quantitative estimate of past climatic variations independent of historical sources. These reconstructions can be used to develop and test hypotheses concerning the effects of climatic variation and change on man's history. Both data sources provide clues as to possible causal factors and suggest possible human responses to future climatic variations. A closer association and cooperation among historians and dendroclimatologists could provide a well-dated and reliable record and a better understanding of the consequences of these short-term climatic variations.

Alexander T. Wilson

Isotope Evidence for Past Climatic and Environmental Change

Human history as a subject has been developed largely by people who have used written records. The availability of written documents falls off sharply with time and in many areas of the world the written records span only a very short period. This article reviews information of interest to historians which is potentially available in other records—for example, in cave formations, tree rings, and ice sheets—and, in particular, describes the contribution that isotopic chemistry can make to the recovery of such information. The availability and quality of isotopic material does not fall off so sharply with time as does the written record.

In nature many systems are laid down layer by layer, as in the deposition of the material in ocean and lake sediments, the wood in tree rings, the ice in ice sheets, and the calcite on limestone formations in caves (speleothems). In many cases extremely detailed climatic and environmental information is recorded from the time that the particular layer was laid down—and is available for interpretation if we can discover how to "read" the record. The information is stored in the chemical composition or in the isotope ratios of the atoms making up the chemicals which form the deposit. The chemical and isotope data may be expected to provide information on temperature, windiness, and the amount and kind of dust in the atmosphere at the time that the deposit was laid down. It is even possible to obtain information, such as records of man's impact on his environment, which has not been recorded in the written record.

All chemical materials are made up of atoms. Water, for example, is made of atoms of the two elements, hydrogen and oxygen. Each molecule of water is made from two atoms of hydrogen bonded through an atom of oxygen. All atoms of an element have essentially the same chemistry, but they sometimes have a variety of different masses (isotopes). For example, there are two stable isotopes of hydrogen of weights 1 and 2 (written respectively ^{1}H and ^{2}H) and there are three stable isotopes of

Alexander T. Wilson is Professor of Chemistry at the University of Waikato, New Zealand. He is currently Director of Research for the Duval Corporation, Tucson.

oxygen of masses 16, 17, and 18 (written respectively ^{16}O, ^{17}O, and ^{18}O). In each of these cases the lighter isotope (^{1}H and ^{16}O) is by far the most abundant in nature, so that most water molecules are $^{1}H^{16}O^{1}H$, although there are small amounts of $^{1}H^{16}O^{2}H$ and $^{1}H^{18}O^{1}H$. How and why the ratios vary in nature forms a large part of the subject of isotope geochemistry. Some of the findings of isotope geochemistry provide tools and techniques for providing data of interest to historians.

The first application of isotope geochemistry to the determination of past environmental temperatures was proposed by Urey. The basis of his paleothermometer was the distribution of oxygen isotopes between the calcium carbonate laid down on the shells of the marine organisms and the water in which the organism was growing.[1]

This technique has been employed to determine the past temperatures of the oceans of the world by Emiliani and others using the shells of foraminifera (forams). These organisms have shells about the size of a pinhead. In practice, cores of bottom sediment are taken from the ocean bed and the forams are separated. They are then dissolved in pure phosphoric acid and the carbon dioxide which is evolved is used for measurement in a suitable mass spectrometer. Unfortunately, because of the very slow rate of deposition of deep sea sediments and the activities of bottom living organisms, the time resolution of such records is no better than a few thousand years, so that these data are of interest only to historians who study early man.[2]

In addition to a high time resolution, a technique is required which can measure temperature very accurately. It is not generally appreciated just how susceptible a culture is to a small change in temperature. Very small changes in the mean annual temperature of an area can have dramatic effects on agriculture and hence on the human population. During the glacial periods the world cooled by about 6°C. This cooling resulted in most of Canada being covered by an ice sheet which advanced over the Great Lakes as far as Chicago and down the Atlantic seaboard as far as New York. At the same time, a large part of northern Europe

1 Harold C. Urey, "The Thermodynamic Properties of Isotope Substances," *Journal of the American Chemical Society*, MCMXLVII (1947), 562–581.

2 C. Emiliani, "Pleistocene Temperatures," *Journal of Geology*, LXIII (1955), 538–578.

was covered with an ice sheet which extended over Britain almost as far south as London. Even a 1°C lowering of today's mean temperature would make it difficult to grow cereals in Scotland. During the cold period of the 1690s, the crop failures in Scotland had dire consequences for its population and may have led to its loss of independence shortly thereafter. The length of the growing season is the critical climatic criteria for human survival, particularly at high latitudes. In more benign latitudes, erratic changes in climate may also cause problems for a human agricultural community leading to the planting of inappropriate crops.[3]

SPELEOTHEMS In order to obtain temperature records with greater time resolution, Hendy and Wilson studied the calcium carbonate laid down in cave formations on stalagmites. Cave deposits (speleothems) provide stratigraphy with an inherent time resolution sufficient to be of interest to human historians. The research showed that data for studying short-term temperature fluctuations over thousands of years can be obtained by measuring the $^{18}O/^{16}O$ ratio of suitable stalagmites. This technique should enable a high resolution temperature curve to be produced for many regions of the globe.[4]

The mechanism of the formation of speleothems is shown schematically in Figure 1. Carbon dioxide given off from plant roots dissolves limestone ($CaCO_3$) to form calcium bicarbonate which enters the cave in the groundwater. As the groundwater flows over the surface of the speleothems, carbon dioxide is given off into the cave atmosphere, reversing the reaction so that calcite ($CaCO_3$) is deposited on the surface of the speleothems. If isotopic equilibrium is maintained during the deposition, the isotope distribution between the oxygen isotopes of the calcite and the groundwater records the temperature of the cave.

If it can be assumed that a stalagmite has been laid down in isotopic equilibrium with the water, then the oxygen isotopic ratio of its calcite is controlled by two factors: (1) the isotopic

3 See the argument in Andrew B. Appleby, "Epidemics and Famine in the Little Ice Age," in this issue.

4 C. H. Hendy and Wilson, "Palaeoclimatic Data from Speleothems," *Nature*, CCXIX (1968), 48–51; Wilson, Hendy, and C. P. Reynold, "Short-Term Climate Change and New Zealand Temperatures During the Last Millennium," *Nature*, CCLXXIX (1979), 315–317.

Fig. 1 Mechanism of Formation of Speleothems Shown Schematically.[a]

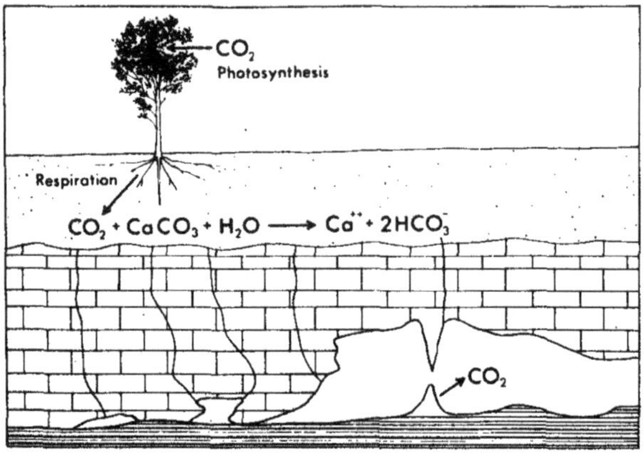

a Carbon dioxide given off from plant roots dissolves limestone ($CaCO_3$) to form calcium bicarbonate (HCO_3^-) which enters the cave dissolved in the groundwater. As the groundwater flows over the surface of the speleothem, carbon dioxide is given off, reversing the reaction so that calcite ($CaCO_3$) is deposited on the surface of the speleothem. If isotopic equilibrium is maintained during the deposition, the isotope distribution between the oxygen isotopes of the calcite and the groundwater records the temperature of the cave.

composition of the water flowing into the cave; and (2) the temperature at which the calcite is deposited on the speleothem, which for many caves can be taken as the mean annual temperature. The isotopic fractionation of the oxygen isotopes between calcite and water will increase by 0.024 percent as the temperature of deposition increases by 1°C. Since existing mass spectrometric techniques enable relative measurements of $^{18}O/^{16}O$ to be made to 0.0025 percent, one should be able to obtain paleotemperature curves to 0.1°C and this should be useful for studying short-term climatic fluctuations.

The $^{18}O/^{16}O$ composition of the atmospheric precipitation percolating into the cave is determined first by the isotopic composition of the oceans, which changes between the glacial and interglacial situation because of the removal of water depleted in ^{18}O and its deposition on the ice sheets. This effect, however, is negligible for the short-term temperature fluctuations under study. The second effect is caused by the temperature difference between the region of evaporation and the area of interest. This effect works in the opposite direction to the effect of temperature and tends to reduce any fluctuations due to temperature. However, if tropical stalagmites are measured, they give temperature changes of the tropics directly. Further, comparisons between stalagmites from tropical and temperate regions give a plot of changes in temperature differences during time, and hence a record of past latitudinal temperature gradients. Such data provide a measure of the intensity of the zonal atmospheric circulation in the past and might throw light on such problems as why the Polynesians stopped sailing and exploring the South Pacific in the fourteenth century, and why, at that time, the Maoris in New Zealand changed their canoe design. (These questions are discussed in more detail below.)

The obtaining of any paleoclimatic data from stratigraphy requires some method of dating. In the case of speleothems, this can be achieved for the last 35,000 years by carbon-14 dating. There is, however, a complication in that the carbon laid down as carbonate on a stalagmite contains a mixture of ancient carbon from the limestone carbonate, which contains essentially no ^{14}C, and modern carbon respired by plant roots. The calcium in solution has been dissolved from the limestone according to the following equation:

$$H_2O + CO_2 + CaCO_3 \rightleftharpoons Ca^{++} + 2HCO_3'$$

Assuming that saturation is reached as the CO_2 solution percolates through the limestone and into the cave, then a little more than half of the carbon in the resultant solution would be of recent biogenic origin and a little less than half would be derived from limestone.

The theoretical situation is not always found as, for example, in the Twin Forks stalagmite, because of the exchange of carbon dioxide with cave air. The details of how such corrections can be made using $^{13}C/^{12}C$ ratios are given by Hendy. In the case of a stalagmite which is still growing when collected, an even more accurate estimate can be made by determining the carbon-14 content of the outer layer of the stalagmite. Other methods, such as thermoluminescence, are also potentially useful for dating material such as calcite. Several samples must be dated across the stalagmite to correct for any variation in growth rate of the stalagmite during the time under investigation.[5]

The Twin Forks stalagmite described in this article came from a cave in northwest Nelson, New Zealand, Latitude 40°40′S, Longitude 172°26′E, and was growing at an altitude of about fifty meters above sea level. The stalagmite was taken near a meteorological station which has a present day mean temperature of 12.2°C. It was sectioned and 30 g samples were taken for ^{14}C determination to provide a time base. Samples (~50 mg) were also taken at regular intervals down the axis of the stalagmite for isotopic analysis with a 1.6 mm steel drill. Five 10 mg aliquots of each of the samples of calcite were reacted with 100 percent phosphoric acid at 25.0 ± 0.1°C in evacuated glass reaction vessels. The carbon dioxide was purified and the ratios of mass 45 to (mass 44 + mass 46) and mass 46 to (mass 44 + mass 45) were compared with a sample of carbon dioxide prepared from Te Kuiti limestone on a Nuclides Analysis Associates 60° double collector mass spectrometer. The $^{18}O/^{16}O$ ratio is reported as $\delta^{18}O$ with respect to the international standard PDB, where

$$\delta^{18}O(^0/oo) = \frac{^{18}O/^{16}O(\text{sample}) - {}^{18}O/^{16}O(\text{PDB})}{^{18}O/^{16}O(\text{PDB})} \times 1{,}000$$

In such a study it is necessary to plot the $^{13}C/^{12}C$ ratio against the $^{18}O/^{16}O$ ratio to ensure that they do not correlate, and hence confirm that the stalagmite was indeed laid down under conditions of isotopic equilibrium. The results of these data are combined as a paleotemperature curve on Figure 2 and are compared with Lamb's curve for central England.[6]

The curve has been scaled in terms of temperature since we know the temperature at which the stalagmite was growing when collected and the temperature change for the last cold period in New Zealand. This cold period occurred during the first decade of this century and was recorded by direct instrumental obser-

5 Hendy, "The Use of Carbon-14 in the Study of Cave Processes," in I. U. Olsson (ed.), *Radiocarbon Variations and Absolute Chronology* (New York, 1970), 419–443.

6 See notes 4 and 5 above. J. M. McCrea, "On the Isotopic Chemistry of Carbonates and a Paleotemperature Scale," *Journal of Chemical Physics*, XVIII (1950), 849–857; H. Craig, "Isotopic Standards, for Carbon and Oxygen and Correction Factors for Mass-spectrometric Analysis of CO_2," *Geochimica et Cosmochimica Acta*, XII (1957), 133–149; Hubert H. Lamb, "The Early Medieval Warm Epoch and Its Sequel," *Palaeogeography, Palaeoclimatology, Palaeoecology*, I (1965), 13–37.

Fig. 2 Speleothem Temperatures from New Zealand to that of Central England.[a]

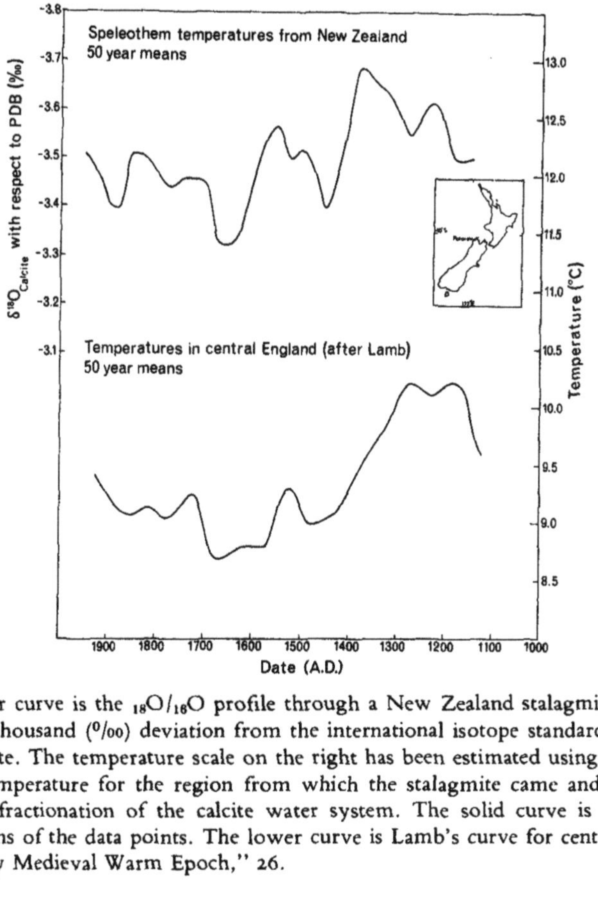

a The upper curve is the $_{18}O/_{16}O$ profile through a New Zealand stalagmite expressed in parts per thousand (^{0}/oo) deviation from the international isotope standard PDB—Pee Dee Belemnite. The temperature scale on the right has been estimated using the modern day mean temperature for the region from which the stalagmite came and the known temperature fractionation of the calcite water system. The solid curve is the 50 year running means of the data points. The lower curve is Lamb's curve for central England. Lamb, "Early Medieval Warm Epoch," 26.

vations. Since the above estimate is in agreement with the known temperature dependence of the fractionation factor of calcite, it suggests that little or no change in the groundwater composition has taken place and that the isotope fluctuations in the stalagmite studied appear to be mainly due to temperature changes.

These results are of a very preliminary nature and the purpose of the analysis is to investigate the potential of stalagmites as indicators of short-term temperature variations. Clearly, many stalagmites should be taken from different caves in different parts of New Zealand. However, it does appear from the above data that the temperature curve for New Zealand is broadly similar to that of England and that such climatic fluctuations as the medieval warm period and the little ice age were indeed of global extent and not just a local European phenomenon.[7]

The comparison in Figure 2 is between a New Zealand curve, where the temperature is known better than the actual date, and the English curve, where the date is known accurately from historical records but the temperature may be in error. However, the last 100 years of the curve agree well with New Zealand meteorological records and not with the English temperature curve during the same period. Whereas the temperature of England since the 1940s has cooled significantly, the mean New Zealand temperature is still climbing.[8]

An interesting feature of the curve is that in recent times samples representing times as short as ten years were measured. Because the hole drilled in the stalagmite is circular in section more calcite is obtained from the central years of the time span with progressively less as the extremes are approached. However, this study shows that the technique is capable of a time resolution of only a few years and is limited only by the winter/summer sampling problem. As can be seen from the New Zealand temperature curve, the very rapid temperature drop in the fourteenth century would have had a catastrophic effect on the Polynesian agriculture—based as it was on the tropical food plants, taro and kumera—in all except the very north of the North Island, and it is a tribute to Maori agriculture that these tropical food plants survived at all. It is probably no coincidence that the Polynesian exploration of the South Pacific came to an end at this time. A cooling might have been expected to lead to steeper latitudinal temperature gradients and hence to a more violent climate. The loss of travellers because of the increase in frequency of more violent storms would have effectively discouraged further exploration.[9]

DETERMINATION OF PAST WINDINESS The way in which man spread across the Pacific is difficult to understand. The dispersion

7 But see Reid A. Bryson and Christine Padoch, "On the Climates of History," in this issue.

8 K. E. Trenberth, *Climate and Climatic Change: A New Zealand Perspective* (Wellington, 1977); Lamb, "The New Look of Climatology," *Nature*, CCXXIII (1969), 1209–1215.

9 Wilson and Hendy, "Past Wind Strength from Isotope Studies," *Nature*, CCXXXIV (1971), 344–345.

appears to have taken place during periods of intense activity interspersed with periods of little exploration. The last period of rapid expansion was between 900 and 1350, during which time a significant part of the South Pacific was settled. This period, interestingly enough, began and ended at the same time as the Viking colonization of the North Atlantic. Could it have been the increase in storminess with the onset of a cold period that led the Vikings to abandon their North Atlantic route rather than seek a more southern route? Similarly, the Polynesian migrations in the Pacific started with the onset of a warm period and ended at the onset of the little ice age.

It has always been difficult to understand how the Polynesians could have settled New Zealand from Tahiti by the early fourteenth century and then not have gone on to discover Tasmania and Australia. If the medieval warm period had been a time when very violent storms were rare, this would explain how the Polynesians were able to explore and settle the South Pacific in a few hundred years. If the break in climate in the fourteenth century caused a sharp increase in the frequency of violent storms, the loss of a relatively few skilled navigators could have effectively discouraged further explorations.

It can be seen from these questions that in order to understand the history of certain areas of the world, such as the South Pacific, it would be useful to have a record of the past zonal atmospheric circulation—that is of past windiness. To discuss how this might be done, we need to review isotopic meteorology and, in particular, the meteorological data contained in the polar ice caps.[10]

The atmosphere of the earth is in effect a heat engine, the heat source being at the earth's surface at low latitudes and the heat sink being in the polar regions. The temperature difference between the equatorial and polar regions, which we will call ΔT, largely determines the intensity of the zonal atmospheric circulation. If we could determine how ΔT has varied over past periods we would have a measure of how the strength of zonal winds and storminess (i.e. more intense and/or more frequent depressions) has varied over that time. As will be shown below, this information is recorded in the $^{18}O/^{16}O$ and $^{2}H/^{1}H$ ratios of the ice in the Greenland and Antarctic ice caps.

10 Wilson and Hendy, "Climatic Implications of the Isotope Ratio Profiles through the Polar Ice Caps," in E. M. van Zinderen Bakker, Sr. (ed.), *Palaeoecology of Africa: The Surrounding Islands and Antarctica* (Cape Town, 1973), VIII, 117–124.

The fractionation of ^{2}H and ^{18}O in atmospheric precipitation has been worked out by Friedman et al. and Dansgaard. It was found experimentally that as one moves to successively colder regions, precipitation in the form of rain or snow becomes progressively depleted in ^{18}O. This depletion is proportional to the temperature difference between the point of evaporation and the point of condensation.[11]

Water in the hydrosphere, which is made up of the lighter isotopes of hydrogen and oxygen, condenses at a slightly lower temperature than water with one of the heavier isotopes, either ^{18}O or ^{2}H. Thus there is a progressive depletion in the heavier isotope in rain, or snow, as one moves away from the equator (or up a mountain). Because there is more evaporation in the equatorial regions than precipitation and the reverse in the higher latitudes, water vapor is always being removed selectively. Theory predicts that the amount of depletion will depend on the temperature difference (Δt) between the point of evaporation and the point of condensation. Thus, if one measures the $^{18}O/^{16}O$ ratio of precipitation at a point, one can determine the temperature difference between that point and the equatorial region. In studies of atmospheric precipitation, exactly equivalent data are obtained from both $^{18}O/^{16}O$ studies and from $^{2}H/^{1}H$ studies; for simplicity, only $^{18}O/^{16}O$ work will be considered here.

Using the data given by Dansgaard for the isotopic composition of present-day precipitation falling on various parts of the earth, it is possible to prepare a plot of depletion of ^{18}O in precipitation at a given point against the temperature difference between that point and the equatorial regions.[12]

Studying Δt as a function of time for various regions of the earth is in itself an interesting problem which can be studied using data from speleothems or water stored in aquifers since glacial times. However, to obtain a measure of the past violence of the earth's weather as a whole, one needs the temperature difference between the equatorial and polar regions, and hence the $^{18}O/^{16}O$ (or $^{2}H/^{1}H$) ratio of the polar precipitation as a function of time.

11 I. Friedman, A. C. Redfield, B. Schoen, and J. Harris, "The Variation of the Deuterium Content of Natural Waters and the Hydrologic Cycle," *Reviews of Geophysics*, II (1964), 177–224; W. S. Dansgaard, "Stable Isotopes in Precipitation," *Tellus*, XVI (1964), 436–468.

12 *Ibid.*; Wilson and Hendy, "Isotope Profiles through Ice Caps," 117–124.

Recently such data have become available in the form of $^{18}O/^{16}O$ profiles through the Greenland ice sheet at Camp Century (Lat 77°N) and for $^{18}O/^{16}O$ and $^{2}H/^{1}H$ profiles through the Antarctic ice sheet at Byrd Station (Lat 80°S). Thus, it is possible to produce a plot of how ΔT has varied over the 100,000 years for both the northern and southern hemisphere. The most striking result is that during glacial periods the temperature difference between the equatorial and polar regions increased by 20 to 25 percent. Thus, the last glacial period, and by implication earlier glacial periods, were characterized by much more violent weather.[13]

Such studies show that as the world gets cooler the zonal atmospheric circulation increases, the climate becomes more windy, and the frequency and violence of storms increase. This cycle supports the suggestion that it was changes in storminess which ended the migrations of the Polynesians and probably also the Viking explorations.

ISOTOPE STUDIES OF TREE RINGS Tree rings present another exciting possibility for obtaining high resolution paleoclimatic information. The science of dendrochronology can provide sequences of tree rings from various parts of the world. In some places these sequences extend back for more than eight millennia and provide samples of wood accurately dated to one year. The interpretation of isotope data from tree rings (isotope dendroclimatology) is still in its infancy but has the potential of providing detailed information. Isotopic dendroclimatology studies face two problems. First, at what times of the year is the material that is ultimately formed into wood components drawn from the atmosphere? And second, how does the isotopic composition of a wood component, such as cellulose, vary with changing climate?

A conifer manufactures photosynthates at all times of the year except when climatic conditions are unsuitable, due, for example, to low temperatures or drought stress. These photosynthates are stored for short or extended periods before they are laid

13 Dansgaard, S. J. Johnsen, J. Moller, and C. C. Langway, "One Thousand Centuries of Climatic Record from Camp Century on the Greenland Ice Sheet," *Science*, CLXVI (1969), 377–381; S. Epstein, R. P. Sharp, and A. J. Gow, "Antarctic Ice Sheet: Stable Isotope Analyses of Byrd Station Cores and Interhemispheric Climatic Implications," *Science*, CLXVIII (1970), 1570–1572; Wilson and Hendy, "Isotope Profiles through Ice Caps," 117–124.

down as wood. Most conifers lay down wood only during a brief period of the year. The actual period of wood deposition is not controlled by net photosynthesis but is under hormone control, the production of hormones being controlled principally by day length and, to a lesser extent, by temperature. For example, in trees suffering from drought stress, wood can be laid down during periods when net photosynthesis is negative.[14]

The above considerations apply to any isotopic work on trees. But in the case of isotopic work on oxygen and hydrogen another factor is important. The isotopic composition of rain or snow depends on many factors including the temperature history of the air masses which bring the precipitation to an area. This is not the end of the problem, however. Once the water is taken into the tree, transpiration processes in the leaves can cause large fractionations particularly in arid environments.[15]

Under favorable situations, the $^{13}C/^{12}C$ ratio of both cellulose and lignin can be used to determine past temperatures. There are good prospects for using the D/H and $^{18}O/^{16}O$ ratios to determine past aridity and even past weather patterns. In Colorado, for example, the storm tracks can come over the mountains from the Pacific, in which case the water they contain is depleted in the heavy isotopes. Alternatively, they may come up from the Caribbean, in which case they are much less depleted. Since trees record the precipitation, one might hope to recover past climatic patterns from isotope studies on tree rings. The problems in the interpretation of the isotope data from tree rings are formidable but the prize is a time resolution better than one year and an absolute dating system accurate to one year.[16]

Isotope dendroclimatology can never be expected to produce a mean annual temperature curve as might be produced by a meteorological observer or from measurements of speleothems. It will produce only a curve representative of some period of the year, for example, a spring and early summer temperature curve. Different trees may draw from the atmosphere the carbon which is ultimately formed into wood at different periods of the year. A deciduous tree, such as an oak, carries out photosynthesis in

14 Harold C. Fritts, *Tree Rings and Climate* (London, 1976).

15 Friedman et al., "Deuterium Content of Natural Waters," 177–224.

16 Wilson and M. J. Grinsted, "$^{12}C/^{13}C$ in Cellulose and Lignin as Palaeothermometers," *Nature*, CCLXV (1977), 133–135.

the spring and summer, whereas at the same site a conifer, such as *Pinus radiata,* can draw carbon from the atmosphere throughout the year. Thus, the wealth of paleoclimatic data recorded in the isotopic ratios of the constituents of tree rings and the superb time base provided by dendrochronology makes the potential of isotopic dendroclimatology great.[17]

ENVIRONMENTAL EFFECTS Tree rings, ice sheets, and other stratigraphic sequences are even today recording mankind's activities, sometimes in a more quantitative form than is recorded by man himself. The development of the automobile is recorded in the increased lead (from the tetraethyllead in the automobile fuel) content of the Greenland ice sheet.

An interesting example of man's impact on his environment is recorded in the tree rings of bristlecone pine (*Pinus longaeva*). The information gathered from the isotope data in the tree rings described below shows that during the latter half of the nineteenth century, there was rapid development of agriculture in the new lands of North America, New Zealand, Australia, South Africa, and in Eastern Europe. This pioneer revolution had hardly begun in 1850, yet large areas had been cleared and/or plowed by 1900. The change was almost globally synchronous and must have led to the release into the atmosphere of large quantities of carbon dioxide (CO_2). If we consider that a standing forest may contain up to 30,000 metric tons of carbon per km^2 in the standing wood, and a smaller, but still significant, amount in the litter and soil organic matter, much of the carbon is rapidly lost to the atmosphere once the forest is removed. Similarly, virgin grassland soils have a very high organic matter content of up to 5,000 metric tons km^{-2}, half of which may be rapidly lost as a consequence of cultivation.[18]

An order of magnitude calculation shows that the total carbon released into the atmosphere by man's agricultural practices might have been of comparable magnitude to that released by the consumption of fossil fuel. Although historical records exist on

17 Wilson and Grinsted, "The Possibilities of Deriving Past Climate Information from Stable Isotope Studies on Tree Rings," *DSIR Bulletin* [New Zealand Department of Scientific and Industrial Research], 220 (1978), 61–66.

18 Wilson, "Pioneer Agriculture Explosion and CO_2 Levels in the Atmosphere," *Nature,* CCLXXIII (1978), 40–41.

the development of agriculture, details such as the chemical composition of the virgin soil are not known accurately. Here we determine independently how much carbon dioxide (CO_2) was added to the atmosphere from the biosphere, and over what periods.[19]

Because fossil fuels, such as oil and coal, have ages much greater than the half life of ^{14}C (5,730yr) they contain essentially no ^{14}C. Thus the $^{14}C/^{12}C$ ratio of the atmosphere provides us with a measure of how much and for how long a $^{14}CO_2$ molecule injected into the atmosphere remains in the gas phase before it is absorbed or exchanged into the ocean. The change in the $^{14}C/^{12}C$ ratio of the atmosphere during the industrial revolution is called the industrial or Suess effect, and has been determined by measuring the ^{14}C content of tree rings. The total amount of ^{14}C-free CO_2 injected into the atmosphere due to the combustion of fossil fuel until 1950 was equal to 9 percent of the amount already in the atmosphere. On a simple mixing model one would expect the specific activity of the atmosphere to have been reduced by 8 percent. In fact, it was reduced by only one quarter of this amount as a result of isotope exchange and absorption into the oceans. Of the CO_2 currently being produced by the combustion of fossil fuel, half is being rapidly absorbed by the ocean, so that for each metric ton of carbon released into the atmosphere as CO_2, the carbon inventory of the atmosphere rises by only 0.5 metric tons. These two facts enable us to compute the fate of the CO_2 injected into the atmosphere by the pioneer agriculturalists.

Standing trees and soil organic matter contain "modern" $^{14}C/^{12}C$ ratios. (Organic matter that is eighty years old is depleted in ^{14}C by only 1 percent.) However, because photosynthesis fractionates strongly against ^{13}C, trees, soil organic matter, peat, coal, and petroleum have $^{13}C/^{12}C$ ratios depleted by about 2.5 percent (usually expressed −25 per mille) with respect to the international isotope standard PDB—Pee Dee Belemnite. The CO_2 in the atmosphere, however, is only depleted by 0.7 percent (or −7 per mille PDB). Thus, if we could determine the $^{13}C/^{12}C$ ratio of the atmosphere in the past, we could determine how much CO_2 has

19 B. Bolin, "Changes of Land Biota and Their Importance for the Carbon Cycle," *Science*, CXCVI (1977), 613–615.

been released due to the clearing of forests, the draining of peat land, and the plowing of virgin grassland.

One approach would be to measure the $^{13}C/^{12}C$ ratio of the wood of tree rings and this has been tried by several workers, not in order to study the pioneer agriculture effect but to obtain a measurement of the industrial effect independent of that obtained by ^{14}C measurements. Unfortunately, this attempt has not proved fruitful because it has not been appreciated what factors, other than the isotopic composition of the atmosphere, control the isotope composition of the carbon in the wood laid down by trees.[20]

It is only recently that a serious attempt has been made to study what factors control the fractionation of carbon isotopes by trees. The research has been undertaken by workers in the embryonic sciences of isotope dendroclimatology in order to obtain past climate information from the isotopic ratios of the hydrogen, oxygen, and carbon isotope ratios in the various constituents of wood. Briefly, the $^{13}C/^{12}C$ ratio of a plant constituent, such as cellulose, reflects the $^{13}C/^{12}C$ ratio of the atmosphere, but there is a temperature coefficient of 0.02 percent per °C. This coefficient is small enough to be of little consequence in the investigation. It is important to ensure that the trees are, in fact, photosynthesizing in air representative of the atmosphere, and not in air contaminated by cities or industrial operations. An even more serious problem arises because the atmosphere in closed canopy forests is seriously contaminated isotopically by CO_2 respired from the roots of the trees. This CO_2 is made from sucrose translocated into the roots from the leaves, and is isotopically depleted (−18 per mille) in ^{13}C with respect to the free atmosphere.[21]

In order to run an accurate experiment, wood from a bristlecone pine tree was used. This particular tree grew on the lower forest border in the White Mountains of California at an altitude

20 J. G. Farmer and M. S. Baxter, "Atmospheric Carbon Dioxide Levels as Indicated by the Stable Isotope Record in Wood," *Nature*, CCXLVII (1974), 273–275; H. D. Freyer and L. Weisberg, "Dendrochronology and ^{13}C Content in Atmospheric CO_2," *Nature*, CCLII (1974), 757.

21 C. D. Keeling, "The Concentration and Isotopic Abundances of Atmospheric Carbon Dioxide in Rural Areas," *Geochimica et Cosmochimica Acta*, XIII (1958), 322–334.

of 4,000 meters. Bristlecone pine trees grow as isolated trees exposed to the free atmosphere, and lay down wood for only six weeks during the height of summer each year. The wood was carefully cut into samples which had grown in the years of interest. The samples were ground, resins extracted with solvent, and cellulose was prepared. The cellulose was then combusted to CO_2 and measured in a micromass 602C mass spectrometer to a precision of 0.04 per mille (1σ).

The results are presented in Figure 3. The changes in the $^{13}C/^{12}C$ ratio is dramatic from 1870 onwards and is vastly in excess of any possible climatically induced changes. The only explanation is that large quantities of isotopically light CO_2 were liberated into the atmosphere from 1860 to 1890, before the fossil fuel effect really made a significant contribution (see fossil fuel curve in Fig. 3).[22]

The data presented in Figure 3 suggest that the pioneer agricultural effect depleted the $\delta^{13}C$ value of the atmosphere by 0.9 percent. This effect would have been four times as great, or 3.6 percent, if we had corrected for the reduction due to the exchange and absorption into the oceans. Such a change would have been caused by the addition of 110×10^9 metric tons of carbon to an atmosphere which had a δ^{13} value of -6 percent and already contained 550×10^9 metric tons of carbon. Thus the pioneer agricultural effect in the brief period between 1860 and 1890 contributed one and a half times the amount of CO_2 produced by all the fossil fuels burnt up to 1950 (60×10^9 metric tons carbon). The results presented in Figure 3 suggest that the pre-1850 atmosphere had a $\delta^{13}C$ value of -6 percent with respect to PDB and a concentration of about 270 p.p.m. CO_2, rather than the 300 p.p.m. usually assumed.[23]

These data also suggest that, as a result of the pioneer agricultural effect, the CO_2 concentration of the earth's atmosphere changed suddenly over a few decades in the late nineteenth century by about 10 percent [$1/2$ of $(110/550) \times 100$]. This effect was reinforced and increased during the twentieth century by the

22 Wilson and Grinsted, "$^{12}C/^{13}C$ in Cellulose and Lignin," 133–135, Keeling, "Industrial Production of Carbon Dioxide from Fossil Fuels and Limestone," *Tellus,* XXV (1973), 174–198.

23 G. S. Callendar, "On the Amount of Carbon Dioxide in the Atmosphere," *Tellus,* X (1958), 243–248.

Fig. 3 Pioneer Agriculture Effect.[a]

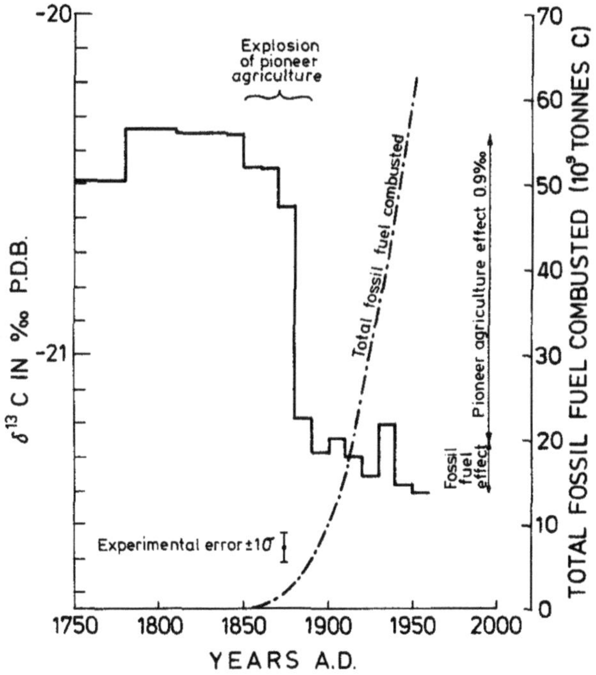

a Plot of δ^{13}C values of the cellulose across a section of Bristlecone Pine. Note the sharp change between 1860 and 1890 which occurred before the combustion of fossil fuel had contributed significant quantities of CO_2 to the atmosphere. As discussed in the text, this could only have been caused by the activities of the pioneer agriculturalists.

growing use of fossil fuels, so that currently the CO_2 levels in the atmosphere have been raised to 330 p.p.m., that is, 22 percent of the pre-1850 levels.

What effect might this increase in CO_2 levels have had on the climate of our planet? On many climatic models an increase of 10 percent in the CO_2 levels in the atmosphere would be expected to raise the world temperature by several tenths of a °C. It is generally accepted that until the middle of the nineteenth century the earth was in the cold period known as the little ice age. Although there are shorter term climatic oscillations within a period of decades, the general level of temperatures during the little ice age was perhaps 0.5°C lower than, say, the first half of the twentieth century. The 10 percent increase in the CO_2 levels of the atmosphere, due to the pioneer agricultural effect, which was later reinforced in the twentieth century by the fossil fuel effect, provided a mechanism whereby the world's temperatures were raised by 0.5°C in the late nineteenth and early twentieth centuries, and the earth emerged from the little ice age.[24]

These results have important implications for agricultural historians. There was a period of not longer than three decades in the latter part of the nineteenth century when there was a much greater expansion of agriculture than during any period before or since. This expansion may have followed the development of railroads, which opened up land not previously accessible. The pioneer agricultural explosion, which occurred more or less simultaneously in very different parts of the world—North America, Eastern Europe, New Zealand, Australia, and South Africa—and began in the 1870s, would be the first and perhaps the most significant of mankind's global assaults on our environment.

This article shows that records of interest to historians exist not only in written records but also in the chemical and isotope composition of stratigraphic deposits in cave, ocean, and lake sediments, tree rings, and ice sheets. Even today stratigraphic sequences may be recording some aspects of man's activity more accurately and more quantitatively than man himself.

24 S. Manabe and R. T. Wetherald, "The Effects of Doubling the CO_2 Concentration on the Climate of a General Circulation Model," *Journal of Atmospheric Science,* XXXII (1975), 3–15; S. H. Schneider, "On the Carbon Dioxide-Climate Confusion," *Journal of Atmospheric Science,* XXXII (1975), 2060–2068; Lamb, "New Look of Climatology," 1209–1215.

Donald G. Baker

Botanical and Chemical Evidence of Climatic Change: A Comment

To the surprise of many, the influence of climate is frequently neither obvious nor easy to detect in historical and cultural records. It is evident that catastrophic climatic events do play important and recognized roles, yet even in certain climatic catastrophes the effect is not the result of climate alone, but is in part a function of man's action as well. Two such events of recent memory are the droughts in the Sahel region of Africa and the Great Plains' "Dust Bowl" of the 1930s. In both events man's activities intensified and worsened the climatic effects. In most cases, however, the climatic effects are modified or diffused to the point where they become difficult to identify with certainty.

There are three reasons why the biological systems upon which climate is acting may not show direct and obvious responses, thus making the historical record difficult to interpret. One reason rests with the biological systems themselves, one is external to them, and one is a matter of scale.

The resiliency that a biological system may exhibit toward a change in the climate can be remarkable. This ability to rebound following an imposed external stress is due to both physical and physiological characteristics of the system. A simple example is the rolling of maize leaves during periods of moisture stress, which serves to reduce leaf transpiration.

External to the biological system are economic, political, and social structures that man has created. They often decrease our ability to detect the effects of climatic events. Commerce provides an excellent example of how the effect of even a climatic disaster in a region or country can be dissipated. Another example is the shifting by farmers from one crop to another in response to either climate or economics—which may be different faces of the same coin.

A third reason for an apparent decreased effect of climate upon the historical record is the temporal and areal scale that is considered. The internal and external factors just noted are such

Donald G. Baker is Professor in the Department of Soil Science at the University of Minnesota.

that only on a microscale record will the climatic effect be consistently evident. On other scales the effects are frequently hidden or masked by seemingly conflicting reactions and interactions.

In spite of the problems noted, climate should never be overlooked as an important influence upon the directions that history and cultures take. It should be recognized that, on most occasions, by the time that the effects have reached the historical record they may be both subtle and indirect.

Historical weather records are all too brief, and both the historian and climatologist have reason to extend the records as far back in time as possible. The unavailability of reliable historical records means that substitute evidence must be found. The substitute or proxy data may be a written record of some kind, but for any extended period of time a biological or geological record must be used. However, with the use of such materials it should not be forgotten that they remain as proxies or substitutes for the desired data.

Thompson Webb, Harold C. Fritts et al., and Alexander T. Wilson describe extremely interesting methods of arriving at estimates of past climates. Two of their methods are similar in that the yield of biological material, pollen in one case and wood in the other, is used to recover the climate at the time that the material was produced. This is a reversal of a familiar technique in agricultural circles where climatic data are used to predict grain yield. Wilson introduces a novel method in which the ratio of ^{18}O to ^{16}O in cave deposits and in other natural systems, such as lake sediments and ice sheets, is used to estimate temperatures at the times of deposition.

All three methods deserve attention and all three can contribute to the knowledge of past climates. However, in our attempt to extend the climatic record we must not overlook certain inherent problems which may limit the ability of each method to provide the kind of quantitative results often asserted for them.

Webb's well-organized and thoughtful introduction to the topic of pollen analysis reminds us of the care which must be exercised in a climatic analysis based upon biological material. He lists three critical assumptions. The first is that there have been no evolutionary changes during the study period—a matter of some concern when dealing in terms of thousands of years. Although this assumption is generally accepted, there remain some nagging questions. For example: a question of the constancy of the genetic makeup of the material can be raised based upon the latitudinal migration of vegetation that must be assumed as pollen assemblages change from time to time in response to the inferred climatic changes. Can such a latitudinal (north-south) change occur without a change in the photoperiodic response of the plant? Such an important variation requires an evolutionary change in the biological material. Another example: some pollen assemblages that have been identified have no modern analogs. Is this indicative of an error in the technique or were there, in fact, different vegetation combinations? If the latter is true, this may be the result of genetic changes in the plants.

Webb's second and third assumptions apply equally to the method discussed by Fritts et al. The assumption of the adequacy of modern data is an important one, and I do not question the acceptance of this assumption simply on the basis of some unobtainable idealistic goal with respect to the quality and distribution of climatic and biological data. But the fact is that today's meteorological data regrettably do not have the accuracy or quality control often assumed. For example, in the month of March the Indiana state-wide mean temperature is calculated to have decreased 0.7°C(1.2°F) in the last forty years. Yet this climatic change occurred simply as the result of a general shift by the National Weather Service cooperative observers in Indiana from afternoon to morning observations. It may be assumed that similar changes in the basic weather data have occurred elsewhere. The result is that much of our "basic" data rest on shaky foundations.[1]

A glance at Figure 4 of Webb's article shows that the temperature requirement of the oak is poorly known—certainly not with the precision required for anything except rough estimates of the temperature climate of the vegetation combinations in which oak, for example, is now found. Figure 4 indicates that retrospective predictive power exists only when oak pollen is less than 30 percent. The same lack of precision is true for our knowledge of current geographic distributions of the various plant communities and the climates associated with each.

1 Larry A. Schaal and Robert F. Dale, "Time of Observation Temperature Bias and "Climatic Change," *Journal of Applied Meteorology*, XVI (1977), 215–222.

There is no question but that certain biological responses are related to physical characteristics of the environment, which is the third assumption. However, I cannot accept the precision indicated by both Webb and Fritts et al. I have two reasons. First, air temperature and precipitation, the climatological elements predicted from the measured biological response, are not primary factors when it comes to controlling biological response. They are of secondary or tertiary importance in the hierarchy of atmospheric parameters. For example, it is apparent that the amount of water that enters the soil and actually becomes available to the plant is a highly variable fraction of the measured precipitation. A closer relationship does exist between plant (leaf) temperature and air temperature than between precipitation and plant-available soil water. Air temperature and precipitation are used by most investigators of yield-climate relationships not because these parameters are of first-order importance, but because they are the most universally measured of the climatological elements.

A second reason for my believing that the indicated precision is questionable is the method used by Webb and Fritts et al. to develop the yield-climate prediction models. Both authors prefer to use deterministic models to establish the yield-climate relationship and thus, in the words of Webb, "the associated shortcomings [of empirical procedures] can be avoided." However, operational deterministic models simply are not available; nor are the necessary data for their use. For purposes of expediency an empirical method must be used. But herein lies a major problem, as the following example will show. A program in the federal government attempts to predict world grain yields from climatic data, a reversal of the Webb and Fritts et al. models, which reconstruct climate from yield data. It was found that, even with the use of current meteorological data, more than forty separate empirical equations are required to predict the Russian wheat yield. (It remains uncertain whether these equations will satisfy the goal of even predicting 90 percent of the yield in nine years out of ten.) In other words, the equations are site specific and variety specific.[2]

2 Dale E. Phinney, Richard G. Stuff, Michael H. Trenchard, and A. Glenn Houston, "Accuracy and Performance Characteristics of LACIE Yield Estimates," paper delivered at the Conference on Agriculture and Forest Meteorology (Minneapolis, 1979).

The most accurate results in the application of an empirical yield-climate regression equation occur when the independent data are confined to the range of the parameters used to construct the equation. Yield values outside this range rely heavily on extrapolation and the assumption that the regression equation continues to hold. In other words, climates different from those experienced in the last 100 years or so cannot be accurately described unless the above assumption holds. Such a circumstance is of major concern if a climatic change occurred during the prediction period, as happens to be the case.

Referral by Webb to the statistically derived confidence limits is to be complimented, since the reader should keep in mind these important constraints. We all need to be reminded, in addition, that these limits are based on the arbitrary and limited set of data used to develop the regression equations. As a result, these statistically derived confidence limits represent a much narrower band or range of values for a given degree of confidence than is true in nature.

Fritts et al. have provided a great deal of detail and reconstructed temperature and precipitation maps. It should be noted that, although the verification procedure of Fritts et al. is rigorous with respect to determining if the model contains climatic information, it does not provide a measure of the precision that may be expected in the result.

A remarkable climatic record exists in Minnesota, which permits verification of the numerous temperature and precipitation reconstructions shown by Fritts et al. in their Figures 4, 5, and 8, which they indicate are tentative and subject to change as more information becomes available. The Twin Cities (Minneapolis-St. Paul) record began with temperature measurements in October, 1819, and precipitation in 1837. The record portion, which overlaps that of five northern Minnesota bench mark stations, has been verified for all but the most recent portion.[3]

A nearby rural station immediately south of the Twin Cities (Farmington 3 NW), with observations now made by the third generation of the Akin family, permits correction of the Twin

3 Baker, "Weather Trends, Climatic Change and the 1975 Season," unpub. paper (Univ. of Minnesota, 1975).

Cities record for urban and other effects. These corrected data, supplied through the courtesy of B. F. Watson, a consulting meteorologist, are the source of the comparative data in Table 1. Although the comparison shown in Table 1 is against a single station, an admittedly severe test, it is for temporally averaged variations, which should increase the reliability of the climatic reconstruction.

The method described by Wilson seems to offer great promise and is an excellent companion to the other two. The isotopic evidence obtained from natural systems deposited layer by layer may be able to provide remarkable temperature accuracy. Yet, like other indirect measurements of climate, it is not without its share of assumptions and approximations. Within the constraints of time and space, Wilson has brought these to our attention. Other aspects more directly pertaining to the climate include the fact that the generally accepted source-to-site temperature gradient and the associated depletion rate of ^{18}O does not appear to apply to the interior regions of North America. This circumstance may be related to our continental climate, which causes a highly seasonal growth of the speleothems due to unequal precipitation distribution and a reduction or even cessation of seepage water in the winter with frozen soils.[4]

Table 1 Comparison between Actual and Calculated Temperature and Precipitation Departures at Minneapolis-St. Paul for the Indicated Periods.[a]

PERIOD	TEMPERATURE		PRECIPITATION	
	ACTUAL	CALCULATED	ACTUAL	CALCULATED
Spring, 1849 (Fig. 4)	−1.3	0.6	83	<100
Summer, 1849 (Fig. 5)	−1.7	0.3	121	<100
Winter, 1861–1870 (Fig. 8)	−1.2	−1.2	283	90
Spring, 1861–1870 (Fig. 8)	−1.3	0.7	152	82

a The calculated values are estimated from the maps in Figures 4, 5, and 8 of Fritts et al.

4 Russell S. Harmon, Henry P. Schwarcz, and Derek C. Ford, "Stable Isotope Geochemistry of Speleothems and Cave Waters from the Flint Ridge-Mammoth Cave System, Kentucky: Implications for Terrestrial Climate Change During the Period 230,000 to 100,000 Years B.P.," *Journal of Geology*, LXXXVI (1978), 373–384.

The desire to solve the mystery of past climates is a worthy undertaking that deserves reward. But it should not lead one to overlook the assumptions of approximations that have to be made with proxy evidence in order to arrive at a numerical solution. Thus, caution must be advised with respect to the acceptance of quantitative climatic results obtained by these or other methods.

David Hackett Fischer

Climate and History: Priorities for Research

Anyone who studies the literature of climatological history cannot but be astonished by the ingenuity of its methods, and by the sweep of its success. Few fields of historical research are presently in so flourishing a state. During the past several decades climatology has grown into a confederation of many little sciences, each with its own special sources and techniques. There is dendrochronology (the study of tree rings) and palynology (pollen grains); sedimentology (river beds) and stratigraphy (lake bottoms); pedology (soils) and glaciology (ice); lichenometry (algae and fungi) and phenology (the study of recurrent phenomena such as harvests and migrations). And there is historical climatology per se, which reconstructs the record of climate from written materials such as the Norse sagas in Scandinavia, and prayers for rain in Spain, which have been converted into an index of precipitation on the infidel assumption that the more people prayed, the less it rained.[1]

For anyone interested in interdisciplinary history, it is heady stuff—an intellectual event in its own right. But still more interesting than its methods are its empirical findings, which are growing rapidly. Many of the methods of climatology are not new; but they have been newly refined, and are producing results which are increasingly useful to cultural historians.

First, the chronology of climate history has become more accurate than ever before. Techniques of radiocarbon dating, which were at first very crude, have been radically improved in recent applications, and supplemented by other methods such as Potassium-Argon analysis, Thorium-Uranium analysis, the study of oxygen isotopes, and paleomagnetic dating.

Second, climatologists have been increasingly successful in improving the "resolution" of their data. They are able to pinpoint their estimates more exactly, and to organize them in time

David Hackett Fischer is Warren Professor of History at Brandeis University.

1 E. Giralt, "A Correlation of Years, Numbers of Days of Rogation for Rain at Barcelona, and the Price of . . . Wheat," paper presented at the Conference on the Climate of the Eleventh and Sixteenth Centuries (Aspen, 1962); Lange Koch, *The East Greenland Ice* (Copenhagen, 1945).

series which are tighter in their intervals than before. An example is Alexander Wilson's application of isotope chemistry to the study of stalagmites in a New Zealand cave. From that unlikely source, he has succeeded in extracting a series of decennial estimates by a method which was used in the past to generalize about centuries and even millennia. As climatological data grow more precise, they become more applicable to problems of cultural history.

Third, historians of climate are making important progress in reconstructing weather patterns for broad areas of the earth's surface at specific points in time. A case in point is the article in this issue by Harold C. Fritts and his colleagues. They are now able to produce annual, and even seasonal, weather maps for much of North America from 1600 to the present.

Fourth, climatologists have broadened their inquiries beyond the traditional subjects of temperature and precipitation. The articles in this issue discuss many other things—circulation patterns and storm activity, pressure anomalies and solar storms, and much more. Emmanuel Le Roy Ladurie suggested at the conference untapped sources which might even allow someone to study the history of the clouds which have hovered above academic centers since the eighteenth century.

Fifth, other researchers are increasingly engaged in a process of record-linkage which connects scattered bits of historical evidence into coherent general patterns of climate change. One important success of that sort is reported in Helmut E. Landsberg's article, which assembles a set of early instrumental data (recorded on instruments with thirty-six different scales) into a single series for Europe and America in the early modern era.

In all of those five ways, important progress is presently being made. The question, to which this article is a reply, is what priorities might be identified for work in the future. In my judgment, top priority should go to primary research which is directly designed to promote the power of synthesis in climatological history. The sprawling confederation of climatological sciences is presently very strong in its particular parts, but weak in its joints. What is most urgently needed today is a synthesizing venture, conceived on the broadest practicable scale, which might put some of the pieces together in a coherent way.

It might be said that knowledge advances by an alternating course of analysis and synthesis. Historical scholarship sails first on one tack, then the other. In the past several generations, the progress of climatology consisted mostly in the development and application of new analytic tools. Today our needs and opportunities are different. It is a time for synthesis on several different levels: first in the history of climate itself, and then in the study of climate and culture.

The first and most fundamental need is for a synthetic project on the level of empirical description, for everything else must rest upon that base. A descriptive effort of that sort might begin to organize data already at hand, and then find and fill the major gaps. Next it might use established techniques of computer-modelling to create an integrated history of the world's climate as it has changed through time. Finally it could publish the results in a form which would make primary data directly accessible to scholars in other fields.

The work should be carried forward on the broadest possible scale. It should be conceived as the history of world climate; special efforts would be required to collect climatological data in the southern hemisphere and central Asia, where much less is known than in Europe and North America. The project should also be broadly defined in the aspects of climate which it studies—not merely temperature and precipitation, but also electrical activity, solar phenomena, the chemical composition of the atmosphere, pressure anomalies, and circulation patterns. Most important, it should be broadly conceptualized as a synthetic description of a dynamic system, changing through time.

The world's climate has been always in motion, never at rest. Climatic change was always present in the past—but in many different forms. A descriptive history of climate might be conceived primarily in terms of the changing rhythms of climatic change itself. To organize our understanding of that aspect of the subject, we must study not merely the first but also the second derivative of change—the rate of change in rates of change—"deep change" in processes of change themselves.

The second derivative might be used as a way of periodizing the descriptive history of climate in a new way—as a discontinuous series of climatic change-regimes. Periodization is not merely an academic game that historians play. It is the way in which they organize temporal generalizations that gives pattern and meaning to the past. The prevailing periodization of climate history tends to be static rather than dynamic in its nature. It refers to states rather than processes—to periods such as "the little ice age," and the "medieval maximum." It might be useful to think about the history of climate in another way, in which its periods refer not to static conditions but to dynamic processes or change-regimes. The essays in this issue suggest the hypothesis that the main lines might run more or less as follows: first, a glaciating change-regime which began about 20,000 years ago, and ended about 8000 B.C.; then a second change-regime characterized by a different set of rates and rhythms and directions of climate change, from approximately 8000 B.C. to perhaps 1200 A.D.; a third, from 1200 to the mid-nineteenth century; and a fourth from the nineteenth century to our own time.

Somewhere near the end of the third change-regime was the "climate pessimum" which is now commonly called the little ice age. Much of the discussion at the conference which preceded this publication was devoted to the empty question of when the period called the little ice age started and stopped. As long as periodization is conceptualized as a state rather than a process, there is no definitive way of settling such a problem, for a static periodization scheme requires us to draw arbitrary lines across the flow of change. A stronger sort of periodization model might refer to a sequence of different change regimes in the past, which would provide a more empirical and less artificial way of ordering our understanding of climate history.

What I have said refers only to the descriptive history of climate itself. The next priority should go to a second synthesizing task, which might seek to study the connections between climate and culture. On the question of the importance of such a connection, and even of the possibility of finding it, most of the cultural historians who have contributed to this issue are in rather a bearish mood. Le Roy Ladurie has written that "in the long term the human consequences of climate seem to be slight, perhaps negligible, and certainly difficult to detect." That surprising statement was explicitly endorsed at the Climate and History Conference by Andrew B. Appleby, Theodore K. Rabb, and Jan de Vries. The articles by Appleby and de Vries have attempted to establish correlations between climate change and demographic variables, and report generally negative results.[2]

Nevertheless, it appears to me that Le Roy Ladurie's caveat is probably mistaken, and certainly premature. Before we agree that "in the long term the human consequences of climate seem to be slight, perhaps negligible," it might be well to study the question in detail, with that degree of specificity which the progress of historical climatology is beginning to allow.

It is clear that climate and culture have not been connected through history in one simple, universal causal relationship of the sort which so many scholars have tried in vain to discover. But there are other causal possibilities which may be more promising. To pursue them it is necessary to historicize the subject in still another dimension—not merely in respect to climate itself, or culture per se, but also in the history of their conjunction. In the study of climate and culture, the conjunctive term "and" might be understood as a historical variable in its own right. Climate and culture have been connected in different ways, and those connections might themselves be organized into a history.

Consider the following hypothesis. Once, there was a time when variations in climate determined the possibility for human culture to exist at all. That most brutal sort of conjunctive relationship prevailed through most of the earth's history—until perhaps as recently as 10,000 years ago in many parts of the world.

Then there was a second conjunctive period, when climatic change had an effect, not upon the existence of culture itself, but rather on the rise and fall of individual cultures. That was the period which Reid A. Bryson's carbon dates tell us so much about—the period from 8000 B.C. to 1200 A.D., which was so crowded with climatic fluctuations and cultural discontinuities.

A third sort of conjunctive relationship may have developed first in Western Europe, a thousand years after the birth of Christ. This conjunction was another causal connection between climate and culture, in which climatic events had a role in causing crises of subsistence within Western culture without going so far as to threaten its actual existence, as it had done in the case of so many ancient civilizations. It was a time when the Western world was

2 Emmanuel Le Roy Ladurie (trans. Barbara Bray), *Times of Feast, Times of Famine: A History of Climate Since the Year 1000* (Garden City, 1971), 119.

racked by a series of violent disturbances, in which climate played a part, even as the structure of its culture remained in being. Those crises occurred from the fourteenth century to the early nineteenth; the last of them accompanied and followed the Napoleonic Wars. Thereafter, many local crises developed, but none were great or general in the West.

Then there was a fourth period when the conjunction between climate and culture took still another form, which appears in Jerome Namias's article on drought in the twentieth century. Climatic change, now increasingly set in motion by cultural events, caused major alterations in patterns of behavior and interaction within Western cultures without producing major crises. Examples would include adaptive processes such as the great migration from the Dust Bowl to California during the 1930s, or sales of wheat to Russia, with all the political and economic repercussions which that transaction entailed. But even as that fourth conjunctive relationship developed in the West, there were people in other parts of the world whose lives continued to be affected by the third conjunction. A case in point is the tragic difference between the Russian shortages and the African famines which have occurred in our own time. In Africa and Asia, crises of subsistance, climatically determined in some degree, still go on.

If this hypothesis is correct, then the history of climate and culture may be understood as a developing series of conjunctive relationships of great complexity. Within each of those contexts, human beings existed not as mindless automatons—not merely as objects, but also as agents who adapted to environmental conditions and altered them as well, not always as they had intended. From the great land-clearings of the iron age to the impact of modern society on the ozone layer and carbon dioxide levels in the atmosphere, culture has had an impact on climate, as well as the reverse.

In each conjunctive relationship, climate change (caused in part by human acts) created certain challenges in moments which might be called crises of adaptation. We might expect those moments to have occurred when rates of climatic change were unusually rapid, when the magnitude of change was unusually great, when deviations from normal conditions were most prolonged, or when the rhythm of climatic change was most variable and unstable, and one discontinuity was followed so quickly by another that an adaptation to one climatic movement may have exacerbated the effect of the next. Of those many sorts of challenges—changes in the rate, magnitude, duration, and variability of climatic change—perhaps climatic variability (itself conceived as a variable in time) may have been most important.

There are several particularly interesting periods in world history which might be studied with that hypothesis in mind. The first was what Jaspers called the "axial period" of world history, in the sixth and fifth centuries B.C., when so many of the world's great ethical and religious systems were created. Within the span of not much more than a century, we find Confucius and Lao-Tzu in China, Buddha in India, Zoroaster in Iran, the pre-Socratic philosophers in Greece, and the greatest of the old Jewish prophets, who is called Deutero-Isaiah. The ethical systems which those men invented might all be understood as answers to a question, and the question was everywhere the same. How could man create a structure of stable values in a world of unsettling change?[3]

In geological time, the axial age was approximately 2500 B.P., a moment which appears in Bryson's radiometric chronology as a time of extreme and repeated climatic discontinuities—a period of extraordinary variability in the world's weather system. One might ask if there might have been a connection—not in any simple or monistic way, but if Confucianism, Taoism, Buddhism, and Zoroastrianism were each in its own way a cultural response to social disorder caused in part by climatic disturbances.

That is merely one subject which might be studied in detail, before we conclude that the "long term" consequences of climate in human affairs are "slight, perhaps negligible." Another is the rise and fall of the Mycenaean civilization, which almost certainly was climatologically determined in some degree. A third is the coming of the so-called Dark Age in Western history, which appears in Wilson's article to have been a time of extreme climate disturbances.

A fourth is the transatlantic movement of European populations—Scandinavian, Iberian, and West European. There is abundant evidence of the importance of climate as a determinant

3 Karl Jaspers, *The Origin and Goal of History* (London, 1953), 1–21.

of population movements in Scandinavian history—even the most skeptical historians seem willing to agree that the human consequences of climate change were great in "marginal areas" such as the sub-Arctic. The beginning and end of Iberian expansion may also have been influenced in part by climatic events of the sort that Vicens-Vives, the Spanish historian, has suggested in his discussion of the desiccation of the Spanish peninsula.[4]

A fifth important possibility is the role of climatic variations in European and American history from the seventeenth to the early nineteenth centuries. Many of the great social upheavals in that "age of revolution" may have been in part responses to unusually unfavorable and highly variable climatic disturbances which are known to have occurred throughout the Atlantic world at the same time. In all of those instances, the axial period and the dark ages, the era of oceanic expansion and the age of revolutions, important linkages may appear in the relationship between climate and culture—not in the form of a mindless, monistic determinism, but rather in the form of an intricate interaction of challenge and choice.

A third order of priorities might be assigned to the study of the "middle-range" problems which dominate most of the work in social science and social history today. Since Huntington's day, historians and social scientists have generally turned away from questions of climatological determinism. But the Huntington thesis was never really refuted. It was merely ridiculed, because it failed to fit the metaphysical framework of social science in the mid-twentieth century. Perhaps it is time for those issues to be reopened.[5]

A range of questions might be asked about climate and historical demography, climate and historical geography, climate and economics, and much more. A few possibilities might be mentioned by way of example—chief among them is climate and mortality. Appleby has reported the failure of his effort to establish a simple correlation between climate and mortality. But that result may be due not to the weakness of such a connection, but rather to its strength and complexity. A great deal of work is underway on this subject, and we may expect to see from it clear

4 Jaime Vicens-Vives, *An Economic History of Spain* (Princeton, 1969), 13–17.

5 See, e.g., Ellsworth Huntington, *Civilization and Climate* (New Haven, 1915).

evidence that climate made a major difference in respect to the season of death as well as to the magnitude of death rates. Dobson has discovered in her research on mortality in early modern England that temperature and precipitation both had a significant impact upon mortality levels and fluctuations, and I have found a similar pattern in America, where hot summers were much more dangerous than cool ones, and mortality rates varied from the northern to the southern colonies in a manner that must have been climatologically determined.[6]

Climate and regional development must have been connected in many ways in the early modern period. Why, for instance, did American cities grow so much more rapidly in the northern than in the southern states? One important determinant may have been climate, operating upon mortality and in turn upon migration rates. In Boston, during the first quarter of the nineteenth century, the average annual crude death rate was not much above twenty per thousand. In Savannah, during the same period, it was ninety per thousand—and nearly 200 per thousand in an epidemic year. Mortality rates in American cities clearly varied with climate until the public health revolution of the late nineteenth century. Climatological conditions in the American South did not forbid urbanization, but raised its human price to very high levels—higher than many people were willing to pay.

Historians of climate have discovered that climatic differences between the northern and southern parts of the United States may have been significantly greater in the seventeenth and eighteenth centuries than they are today. A study of Atlantic sea temperatures shows that, off the coast of New England, the ocean surface was colder by three degrees Centigrade in the period 1780–1820 than in the early twentieth century—a very large difference. But off the southern coast, the water was warmer two centuries ago. Harold C. Fritts's dendrochronological maps show a pattern which is similar in direction, if not in degree. That work suggests that climatological variations from one American region to another were substantially greater in early American history than in our own time—with important consequences for demographic and economic history.

6 Fischer and Mary Schove Dobson, "The Dying Time," paper presented at the Social Science History Association Meeting (Cambridge, Mass., 1979).

Fig. 1 Sea Surface Temperature Deviations in °C in January: 1780-1820 Values Compared with Modern Normals.

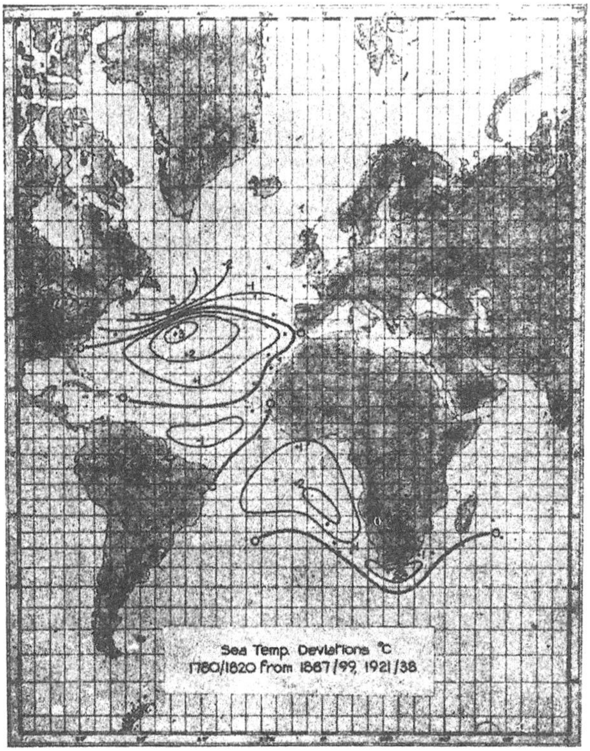

SOURCE: Hubert H. Lamb, *The Changing Climate: Selected Papers* (London, 1966), 16, Fig. 8. Reprinted by permission of the Royal Meteorological Society.

To those brief examples, many others might be added. As our knowledge of climate continues to grow, and as models of the conjunction between climate and culture become more refined, we may expect to see many other interpretative possibilities opening before us. In the meantime, the question of the validity of determinant explanations drawn from climatology puts me in mind of G. K. Chesterton's remark on Christianity. They have not been tried and found wanting, but found difficult and not tried.

Theodore K. Rabb

The Historian and the Climatologist The historian stands amazed before the extraordinary ingenuity of the scientists from half a dozen fields who have manipulated their physical findings to produce detailed information about weather and climate across the centuries. That we can discover the temperature, rainfall, and even windiness hundreds of years ago from such unlikely sources as trees, the sediment on lake floors, lightning strikes in the desert, or cave formations, seems inherently implausible. When one learns about all the possible sources of confusion that have to be filtered out before reliable figures can emerge—the growth pattern or the location of a particular species of tree, the nature of different pollens, or the content of groundwater percolating into a cave—the attainment of results seems almost beyond comprehension.

By comparison, historians have remained relatively close to the inherent meaning of their evidence. They have derived demographic statistics, it is true, from the records of births, marriages, and deaths kept by parish priests, and they have found clues to climate from the dates of wine harvests and the freezing of Dutch canals. But the adoption of powerful physical theory to extract meaning from unlikely data—such as the use of isotope chemistry to recreate wind patterns from the structure of stalagmites—is far beyond their ordinary concerns. These applications of proxy information (for example, extracting variations in world temperature from an ice sheet), are both daunting and fascinating to the toiler who but rarely gets even into the vineyard.

Once the wonder has passed, however, the historian is faced with the same dilemma that arises in the wake of research by demographers, students of popular culture, and other practitioners of recently developed specialties: how does it all relate to the findings of traditional history? In other words, how is one to integrate the new revelations with the familiar political, economic, and social landmarks that historians have painstakingly erected since the generation of Leopold von Ranke? The danger of fragmentation is all too real, with separate sub-disciplines, indifferent

Theodore K. Rabb, co-editor of the *Journal of Interdisciplinary History*, is Professor of History at Princeton University.

to the preoccupations of their next-door neighbors, multiplying the confusion rather than the coherence of our understanding of the past.

The main problems are almost self-evident. First, the time scale of much climatic research is virtually irrelevant to the historian. A few very slow changes, such as the movement of the prairie, can perhaps be linked to discernible events, like the migration of peoples, but the majority of historical research has concentrated on the last 2,000 years, during which 500-year, or even 100-year discriminations have limited value. Second, there is little that historians need to learn when short-term, striking climatic events are self-evidently overwhelming—for example, the droughts and dust bowls of the 1930s. At this level of causation, one can find out all one needs to know from familiar materials: the unmistakable testimony of the millions who saw that their lives were being transformed by the weather. The scarcest, but most necessary type of analysis is the one that lies in an intermediate range between the 500-year world-wide shift, detectable only by climate specialists, and the one-year or decade-long catastrophe, visible to all. Here the possibilities for new insights are endless, arising wherever historians remain unsatisfied about the reasons for major change, and even where they consider themselves satisfied—whether it be the movement northward from the Mediterranean of Europe's economic leadership around 1600, or the end of Polynesian or Chinese voyages more than a century earlier.

A third problem has to do with the chain of reasoning. When studies as disparate as the plotting of sunspot activity and the tracing of volcanic eruptions produce seemingly compatible chronological series, one has to ask whether the results have identifiable consequences or whether, as Bacon observed, "The human understanding is of its own nature prone to suppose the existence of more order and regularity in the world than it finds." Are sunspots and volcanoes independent variables, with traceable effects, or are they both dependent on some *primum mobile*? When he reaches such a cosmic level, the historian is likely to feel safer attributing conjunctions of events to coincidence. Indeed, the determinants of climate, although they form an important subject in themselves, are so distant in the chain of reasoning from what they may ultimately affect in human society, such as the fall of empires, that they seem to belong to an entirely different realm of investigation—that is, unless they themselves include human interventions like deforestation. It is the point at which history and climate *intersect* that offers the most exciting opportunities, but that point is not often reached.[1]

When the meeting of the two does take place, the doubts by no means end. Alexander T. Wilson's fascinating estimate of windiness, for example, adds an important dimension to the weather patterns that are associated with an ice age, and may have a significant bearing on the suspension of Polynesian explorations, as he suggests. But the similar break in the climate in the early sixteenth century had no equivalent effect on Portuguese or Spanish adventurers, and the written evidence does not suggest that European voyages were influenced by the sort of fluctuation in storminess that may have influenced the Polynesians. In other words, the arguments will continue even when climate is brought down to the level of weather, and thus of immediate import in human affairs.

What is essential is that the argument begin on a large enough scale to enrich both climatology and history. Assuming that the technical difficulties of extracting proxy data from the likes of tree rings can be overcome, there is no doubt that information about temperature, rainfall, and wind can give historians new ways of approaching their problems. The suggestion by Harold C. Fritts that the winters of the early seventeenth century were particularly severe leads to all sorts of speculation about the ending of Europe's sixteenth-century economic boom and population rise. Yet Jan de Vries has shown that easy assumptions about the effects of climate have to be tempered by recognition of the varied adaptive capacities of human societies. And both groups of scholars display an urgent need to know much more precisely the nature and the timing of the so-called little ice age, which has been placed (with uncertain consequences and at irregular intervals) anywhere from the fourteenth to the nineteenth century. In sum, the potential for fruitful cooperation remains clear, despite the limited results of the first efforts. Our concern must be to define the best ways of advancing that cooperation in the future.

1 Francis Bacon (ed. Fulton H. Anderson), *The New Organon and Related Writings* (New York, 1960), 50.

Priorities for research are not easy to establish in a decentralized scholarly community. But shared concerns do exist, and attempts can be made to mingle very different types of expertise, if we keep in mind that successful interdisciplinary activity flourishes best when all parties are clear about (1) the context, (2) the method, and (3) the final aim.

THE CONTEXT in this case is an equally powerful interest on both sides—among the historians and among the climatologists. Interdisciplinary work cannot advance very far when one set of scholars merely provides information for another. Joseph Schumpeter is said to have told the historians who enrolled in his seminar that he was delighted to see them, because they could provide the data for his studies of business cycles. Historians have been equally imperious, demanding simple information from psychoanalysts, statisticians, and other colleagues without always engaging in a serious community of effort. This is the "quick fix" mentality—the resort to an alien expertise to solve an immediate problem—and it is rarely effective. There is no doubt that the most useful interdisciplinary research takes place when *both* partners have a stake in the outcome. One of them may have initiated the venture, but unless they share the need for the results, it is unlikely that the mutual benefit will be great.

For historians and climatologists, this context is not difficult to define. They are equally interested in two sets of problems, neither of which can be explored satisfactorily without joint efforts. Both sets of problems, however, are large, providing limitless opportunities for research:

(a) Patterns of adaptation to weather and to climate. This was discussed at some length during the conference, but with only preliminary notions about the substantive issues. Historians want to know if climate "made a difference," and climatologists equally need to discover which of the many phenomena they study have had the largest consequences. Obvious topics include the forms of agriculture (did they change so as to respond to the climate?), the rise of interventionist governments (did they take action to ameliorate natural disasters?), or the emergence of more resilient Europeans who could conquer the world (were they strengthened by having to adapt to a colder climate, and to plague between 1350 and 1750, or were both influences negligible?). To address such questions constructively will take a major commitment from all the relevant disciplines, and this is exactly the context in which the best interdisciplinary undertakings are born.

(b) Man-made climatic changes. Here again there are potential new insights for both the historian and the climatologist. The questions relate primarily to the past two centuries since the beginning of the Industrial Revolution, but not exclusively. Deforestation is not a recent development; nor is the burning of coal in significant quantities. Historians have not often thought to ask how man's activities influence climatic change, but, if the subject were combined with studies of human adaptation, and fostered by a partnership with climatologists, it could prove to be a fascinating new field of research.

These two sets of problems by no means exhaust the possible areas of cooperation, but they do offer examples of the *kind* of topic which lends itself to productive joint effort. Serving the interests of both partners, they provide the appropriate context for interdisciplinary work.

THE METHOD to be followed also requires agreement. I am here thinking not so much of technique as of approach. It is the type of research, rather than the tools to be used, that ought to be mutually acceptable. The appropriateness of particular statistical measures or sociological theories can be left to individual projects to decide, but the nature of the enterprise ought not to be uncertain. Without wishing to legislate on the matter, one can still suggest that, given the state of the field, the most urgent need is for straightforward documentation. The historians and the climatologists are now in much the same situation as Ranke was when he began to study diplomatic history—before they can do anything else, they have to establish a basic chronology.

For this task it is crucial that all possible sources, whether proxy data or manuscript evidence, be combined. Such an enterprise will depend on very close collaboration, because the integration, for example, of tree ring records with testimony about the freezing over of the River Thames cannot proceed without two experts working alongside one another. Not until a reasonably firm and widely accepted chronology of temperature and precipitation levels is established, including annual means, intra-annual variation, and monthly data wherever possible, will it be possible to move ahead to more sophisticated analyses of human adaptability or man-made climatic change. It would be most helpful if geographic variation and other measures, such as windiness, could form part of the chronology, but even global temperature and precipitation data alone would provide the beginnings of a foundation for further research.

Ultimately, one would like to have such information for at least the last 4,000 years, the majority of recorded history. But one has to start somewhere, and a more manageable stretch of time would be advisable. In light of the discussions at the conference, it seems that here, too, there is a classic problem that both historians and climatologists would place near the top of their priorities: the so-called little ice age. Leaving aside the questions of terminology that engender disagreement, and the differences (depending on one's sources) over exact timing, one can detect a broad consensus that somewhere between the fourteenth and the nineteenth century there was a period, or perhaps a number of periods, when temperatures fell, glaciers advanced, and so forth. To determine exactly what did happen—and to see how unusual dislocations, like the Maunder minimum in sunspot activity, or a succession of early grape harvests, fit into a larger picture—a vast collaborative effort is needed. To establish a base line, the collection of data should probably begin in the thirteenth century, and it should probably continue into the twentieth. One can see the information thus gathered ("The Climate of the Little Ice Age," perhaps, or "The 700-Year Climate Survey") forming the basis of radically new research and also serving as a model for such accumulations in other periods of the past. Here, if anywhere, there is a chance for intelligent funding of a well-organized, cooperative project to stimulate major advances across a wide spectrum of scholarship.

THE FINAL AIM of all this work must echo the spirit of genuine interaction with which it is undertaken. The goal has to be no less than the uncovering of a demonstrable pattern of links between climate and political, economic, social, and cultural history. Nobody expects such links to be purely mechanical—along the lines of "a drop in annual mean temperature of *x* degrees causes a population decline of *y* percent, which leads to a reforestation of marginal lands of *z* percent, which in turn leads to a rise in precipitation of . . ." and so forth. Nor are they likely to consist of abstract theoretical statements (which climatologists may prefer) or localized empirical findings (which are historians' forte). A more promising objective would be the development of a special form of analysis. Having examined the effects on one another of climatic change and human action, researchers could strive to establish the *range* within which these two mutual influences operate. It is surely less important to prove that one holds consequences for the other than it is to work out the mechanism and limits of those consequences.

Demonstrating the impact on history of the dust bowls of the 1930s or the terrible winter of 1709 is fairly straightforward, as is tracing the climatic results of air pollution or the reclamation of deserts. Far less easy, and much more important, is the analysis of less obvious cases, involving the middle ground rather than the extremes. Ambiguous situations, in which climatic change or human action is but one of many considerations, offer the best opportunity for careful discrimination among degrees of causation. It is the complicated explanation that demands precision and encourages a definition of the constraints within which influences are felt. Thus, instead of dismissing the relevance of climate out of hand because we know it was not solely responsible for a particular development, we can make it a more integrated part of history by trying to give an appropriate weight to its effects on variables such as the following (to give only a few examples that arose during the conference):

- — the incidence and severity of disease;
- — the price of grains and other foodstuffs;
- — forms of transportation and communication; and
- — the general acceptance of government intervention in daily life.

In none of these areas will climate prove to be the whole, or perhaps even a major part, of the story. Nor will human action alone tell us why climate changed. Large residuals of explanation will always remain. But if we can come to understand more precisely the degree of influence, avoiding blanket generalizations that make climate sound either decisive or insignificant in human affairs, then we can move on to the far more interesting questions of variation, adaptation, and comparisons across time and space.

Emmanuel Le Roy Ladurie and Micheline Baulant

Grape Harvests from the Fifteenth through the Nineteenth Centuries

Grape Harvests from the Fifteenth through the Nineteenth Centuries Several research results have already been published on the subject of meteorological fluctuations during the sixteenth century.[1] In order to interpret the impact of these changes, research based on phenology and in particular on harvest dating continues to be of crucial interest. All other factors being equal, late harvest dates are indicative of a vine-growth period (March-April to September-October) during which average temperatures were mostly cold. Early harvests, on the contrary, indicate relatively high average temperatures during the same seasonal intervals. Temperature fluctuation readings are thus approximations.

This article synthesizes all of the currently known harvest date series (published, or uncovered by us in the French archives) for the vineyards of northern and central France (Paris, Burgundy, and Franche-Comté), Switzerland, Alsace, and the Rhineland. We have omitted the wine regions of southern France to the south of the Geneva parallel and those of western France, west of the Château-du-Loire (Sarthe) meridian, because they fall into another climatic zone.

We have worked out the initial portion of our series in an earlier publication on the sixteenth century (beginning in 1484). For the seventeenth, eighteenth, and nineteenth centuries, we have used 103 local harvest date series from twenty-one current departments in France, French Switzerland, and the southern Rhineland. In all we have grouped together twenty-three units in eleven regions (nine of which are French, one German, and one Swiss). The applied series are taken from early publications by Angot and Duchaussoy, as well as from more recent work. The

Emmanuel Le Roy Ladurie is Professor of History at the Collège de France. Micheline Baulant is a member of the Centre National de la Recherche scientifique.

This article was written with the collaboration of Michel Demonet. An earlier version appeared in *Annales*, XXXIII (1978), 763–771. This is an updated and supplemented version.

1 See for example Le Roy Ladurie (trans. Barbara Bray), *Times of Feast, Times of Famine: A History of Climate Since the Year 1000* (Garden City, 1971).

cut-off date of our research was 1879, the last given date of all the series published in the major article by Angot.[2]

Once these series were recorded on punch cards, the rest of the operation was handled by computer. We chose Dijon as our base series because it had practically no gaps for the entire period under study from 1590 to 1879. (Only four years are missing in the Dijon dossier: 1650, 1794, 1795, and 1814.) Taking into account that the other series were often incomplete, especially in the early periods, we calculated for all the common years between Dijon (d) and each series (s1, s2, s3, etc.) the averge of Dijon (Ad) and the average of the series (As1, As2, etc.). We repeated this operation for each of the centuries studied: the seventeenth century (1590-1790); the eighteenth century (1690-1810); and the nineteenth century (1790-1879). In addition, we used bridge periods to join different centuries (for example, 1690-1710 or 1790-1810) in order to make the comparison and adjustments from one century to the next.

Once we had completed this first phase, it became possible for us to calculate the average difference every century (for each century respectively) between the Dijon series and each of the 102 other local series involved.

$$D1 = Ad - As1; D2 = Ad - As2; \text{etc.}$$

We then added D1, D2, etc., to each year of each series s1, s2, etc., within the framework of the above-mentioned centuries to obtain 102 series, all of which became rigorously comparable to the Dijon series. Using the procedures just mentioned, we eliminated the systematic advance or delay (D1, D2, etc.) in dates of local wine regions in relation to the Dijon wine region which was used as the standard reference.

An example is Salins (Jura) where we have eighty-seven grape harvest dates at our disposal for the nineteenth century (1790-1879) and effectively lack only three years (1794, 1809, and 1843)

2 M. Baulant and E. Le Roy Ladurie, "Les dates de vendanges au XVIe siécle, élaboration d'une série septentrionale," in *Mélanges en l'honneur de Fernand Braudel, Méthodologie de l'Histoire et des Sciences humaines* (Toulouse, 1973). Alfred Angot, "Etude sur les vendanges en France," *Annales du Bureau central météorologique de France* (1885), B29–B120; H. Duchaussoy, "Les bans de vendanges de la région parisienne," *La Météorologie* (March-April, 1934), 111–188. We have also used unpublished series established by Anne-Marie Piuz (Geneva), Baulant (Paris), and Pierre Gombert (Besançon) who kindly provided them to us.

Table 1 Average Annual Date at the Beginning of the Grape Harvest in Northeast France, French Switzerland, and South Rhineland, 1484-1879.[a]

1484	31.22	1550	31.52	1616	6.24	1682	34.73	1748	29.84	1814	37.56
1485	36.72	1551	25.92	1617	30.01	1683	21.30	1749	28.36	1815	27.26
1486	20.22	1552	19.02	1618	35.96	1684	9.24	1750	28.96	1816	54.13
1487	26.22	1553	32.42	1619	25.66	1685	23.70	1751	39.21	1817	40.78
1488	47.22	1554	21.52	1620	31.19	1686	11.14	1752	35.70	1818	23.87
1489	27.72	1555	43.42	1621	46.54	1687	29.57	1753	25.38	1819	28.27
1490	27.22	1556	0.72	1622	27.24	1688	28.80	1754	33.47	1820	36.71
1491	49.72	1557	28.62	1623	22.41	1689	31.33	1755	23.47	1821	45.15
1492	—	1558	25.22	1624	14.56	1690	28.98	1756	37.79	1822	19.92
1493	35.72	1559	7.72	1625	33.93	1691	23.81	1757	28.17	1823	41.56
1494	18.22	1560	32.52	1626	30.00	1692	42.34	1758	27.30	1824	39.45
1495	12.22	1561	20.32	1627	44.64	1693	30.71	1759	24.42	1825	20.80
1496	40.22	1562	25.72	1628	43.11	1694	20.63	1760	22.00	1826	28.68
1497	31.22	1563	30.52	1629	20.10	1695	38.45	1761	21.19	1827	26.51
1498	25.92	1564	37.22	1630	20.25	1696	31.50	1762	17.80	1828	29.88
1499	28.22	1565	32.22	1631	21.41	1697	27.24	1763	36.36	1829	39.93
1500	14.22	1566	25.22	1632	36.42	1698	43.81	1764	23.23	1830	31.36
1501	19.22	1567	21.52	1633	35.71	1699	29.58	1765	31.22	1831	28.45
1502	25.52	1568	33.62	1634	31.03	1700	36.57	1766	30.23	1832	34.65
1503	16.92	1569	31.52	1635	28.40	1701	30.28	1767	42.98	1833	27.92
1504	17.22	1570	38.22	1636	10.91	1702	33.40	1768	32.82	1834	17.87
1505	43.22	1571	14.42	1637	8.73	1703	33.11	1769	32.12	1835	33.71
1506	28.52	1572	21.62	1638	10.43	1704	18.18	1770	43.32	1836	33.06
1507	18.52	1573	42.92	1639	26.02	1705	34.52	1771	33.31	1837	37.38
1508	32.22	1574	32.52	1640	33.30	1706	16.99	1772	30.67	1838	37.07
1509	24.92	1575	26.72	1641	33.03	1707	28.07	1773	38.36	1839	28.06
1510	30.22	1576	34.32	1642	35.37	1708	29.17	1774	29.46	1840	25.50
1511	43.92	1577	32.92	1643	34.84	1709	28.35	1775	29.03	1841	29.36
1512	23.92	1578	24.92	1644	20.18	1710	25.81	1776	34.77	1842	19.26
1513	27.72	1579	39.72	1645	17.42	1711	30.28	1777	37.36	1843	42.28
1514	28.52	1580	28.02	1646	25.56	1712	26.33	1778	25.58	1844	24.56
1515	31.22	1581	38.52	1647	25.98	1713	38.36	1779	22.91	1845	40.70
1516	10.72	1582	29.72	1648	32.70	1714	33.08	1780	23.74	1846	14.79
1517	21.72	1583	14.92	1649	37.25	1715	30.31	1781	13.95	1847	32.94
1518	27.92	1584	24.02	1650	31.83	1716	39.14	1782	35.58	1848	28.27
1519	37.22	1585	36.52	1651	22.01	1717	28.65	1783	21.42	1849	29.07
1520	23.22	1586	32.72	1652	24.91	1718	10.27	1784	19.49	1850	36.74
1521	22.72	1587	38.22	1653	22.40	1719	21.65	1785	30.84	1851	38.22
1522	23.52	1588	27.02	1654	35.09	1720	29.87	1786	33.25	1852	29.27
1523	16.92	1589	26.52	1655	21.47	1721	33.43	1787	38.07	1853	39.28
1524	14.22	1590	11.84	1656	29.51	1722	27.30	1788	17.10	1854	33.74
1525	20.22	1591	31.27	1657	20.99	1723	19.22	1789	36.13	1855	35.81
1526	26.92	1592	34.92	1658	31.51	1724	21.49	1790	30.22	1856	34.70
1527	39.22	1593	33.24	1659	18.00	1725	45.61	1791	22.82	1857	22.13
1528	33.72	1594	34.55	1660	19.00	1726	14.97	1792	33.83	1858	22.33
1529	43.22	1595	31.72	1661	18.81	1727	16.64	1793	26.68	1859	21.08
1530	21.72	1596	35.08	1662	27.47	1728	18.90	1794	14.39	1860	41.84
1531	26.22	1597	42.34	1663	36.05	1729	30.31	1795	28.91	1861	26.19
1532	22.92	1598	28.80	1664	24.28	1730	35.11	1796	35.72	1862	23.94
1533	31.72	1599	12.09	1665	20.77	1731	25.35	1797	30.41	1863	26.81
1534	20.52	1600	42.43	1666	20.48	1732	29.18	1798	20.59	1864	28.53
1535	32.22	1601	41.42	1667	28.88	1733	26.72	1799	43.08	1865	9.41
1536	8.22	1602	21.26	1668	21.49	1734	22.80	1800	25.58	1866	31.84

Table 1 (cont.)

1537 35.22	1603 11.82	1669 18.31	1735 36.71	1801 28.35	1867 29.62
1538 12.22	1604 25.49	1670 23.47	1736 24.02	1802 24.34	1868 13.30
1539 27.92	1605 21.62	1671 19.73	1737 22.60	1803 29.23	1869 24.94
1540 11.92	1606 39.09	1672 29.73	1738 29.19	1804 29.78	1870 17.12
1541 34.22	1607 23.74	1673 39.84	1739 27.64	1805 44.01	1871 31.62
1542 50.22	1608 36.13	1674 26.92	1740 44.32	1806 26.75	1872 30.60
1543 31.22	1609 28.19	1675 48.41	1741 27.74	1807 23.10	1873 30.06
1544 29.72	1610 19.87	1676 11.93	1742 36.20	1808 28.24	1874 21.67
1545 11.92	1611 18.37	1677 29.49	1743 33.18	1809 41.67	1875 25.51
1546 21.72	1612 32.20	1678 23.17	1744 33.54	1810 32.63	1876 31.79
1547 27.22	1613 26.30	1679 25.78	1745 32.64	1811 15.12	1877 32.11
1548 30.72	1614 37.78	1680 16.50	1746 27.58	1812 40.17	1878 33.06
1549 25.22	1615 18.10	1681 19.61	1747 30.40	1813 41.55	1879 46.32

a In number of days counted from September 1 (Gregorian calendar). Although this table first appeared in *Annales*, XXXIII (1978), 765, the present version has corrected numbers and is more up-to-date

out of this ninety-year period. For Dijon, there are also eighty-seven dates for this same time interval (the missing local dates are 1794, 1795, and 1814), so that there are eighty-five common years. On the average during this eighty-five year period the harvest at Salins took place 12.12 days after that of Dijon. The difference, calculated according to the previously mentioned norms, is (under these circumstances) negative: 12.12 days. It is sufficient to add −12.12 days or, what amounts to the same thing, to subtract 12.12 days from each of the Salins figures for the period 1790-1879 in order to obtain a series (for Salins) which becomes, as a result, homogeneous with the Dijon figures.

We then calculated for each of the three major periods or centuries concerned the average by department and then by region; and a general average of all the regional averages for each year. In all we have eleven regions: Côte-d'Or; Jura, Haute-Saône, and Doubs; Cher and Indre; Nièvre and Allier; Marne and Seine-et-Marne; Meuse, Meurthe-et-Moselle, and Vosges; Aube, Yonne, and a part of Loiret; Sarthe, Yonne, and a part of the western part of Loiret; French Switzerland; German Rhineland; and Seine and Seine-et-oise.

Finally, we linked the three periods or centuries. In order to do so we chose as our base the central period, which corresponds to the eighteenth century (1690-1810), and then calculated for the period bridging two centuries (for example 1790-1810, is the bridge period for the eighteenth and nineteenth centuries) the average difference between:

a) all data for 1790–1810 which come from our eighteenth-century series

b) and all data for 1790–1810 which come from our nineteenth-century series.

This slight difference, calculated over twenty-one years, is 0.72 days, which would mean that the nineteenth century apparently had earlier harvests than the eighteenth century. Such a slight deviation can be explained solely by the fact that for each of the two major periods, the eighteenth and the nineteenth centuries, we did not use exactly the same sets of series.

To eliminate this minute difference we added 0.72 to each year of our gross nineteenth-century series (1790-1879), and we obtained a "standardized" nineteenth-century series which is henceforth entirely compatible or "fitted" to the eighteenth century. We carried out the same operation for the gross seventeenth-century series, transforming it into a standardized seventeenth-century series, and for the gross sixteenth-century series,[3] which we transformed into a standardized sixteenth-century series, by comparison (as above) with the seventeenth century, which had been previously standardized according to the above-mentioned norms. In total, we obtained a complete series which goes from 1484 to 1879 for yearly averages, and from 1496 to 1828 for the moving average.

First, we tested this series by internal correlations between regions (see Appendix II). Second, we tested the series by comparison with the Parisian temperature series, from the beginning of April through September inclusive, for the period 1797-1879.[4] Between our global grape harvest series and the temperature series we obtained a correlation coefficient of 0.86 which, given the disparity between the two series (thermometer and grape harvest dates), is excellent and ought to reassure anyone as to the reliability of phenological sources. We did not use the eighteenth-century Parisian series worked out under altogether mediocre

3 M. Baulant and E. Le Roy Ladurie, "Les dates de vendanges au XVIe siècle."

4 Series taken from E. Renou, "Etudes sur le climat de Paris. 3e partie. Température," *Annales du Bureau central météorologique de France* (1889), B195-B226.

Fig. 1 Average Annual Date for the Beginning of the Grape Harvest in Northeast France, French Switzerland, and the South Rhineland Region, 1484–1880[a]

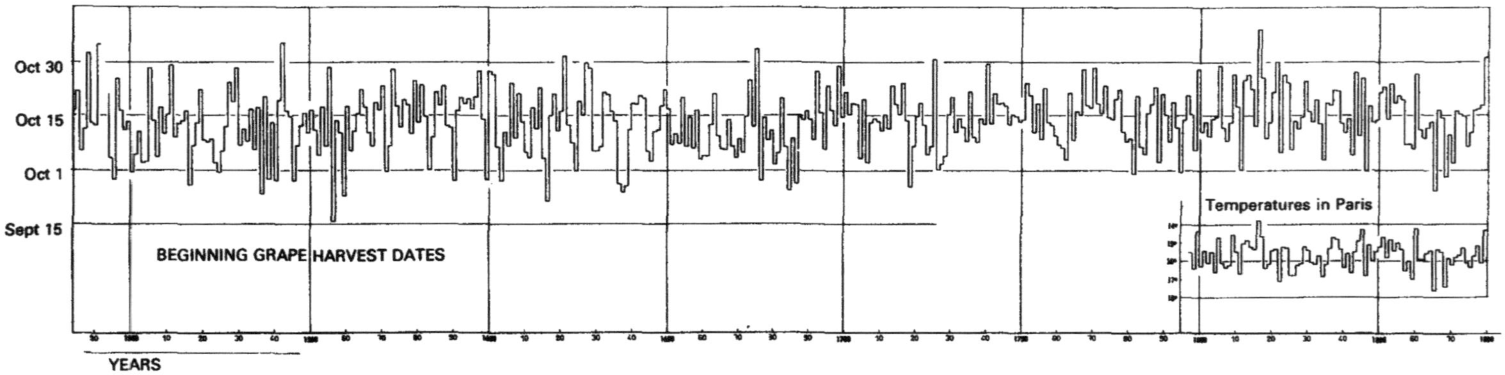

a In number of days counted from September 1. In the lower right corner of the figure are the temperatures in Paris (annual average temperatures) calculated over six months from April to September inclusive.

Fig. 2 Same Series as Figure 1 but tempered by a triennial moving average.

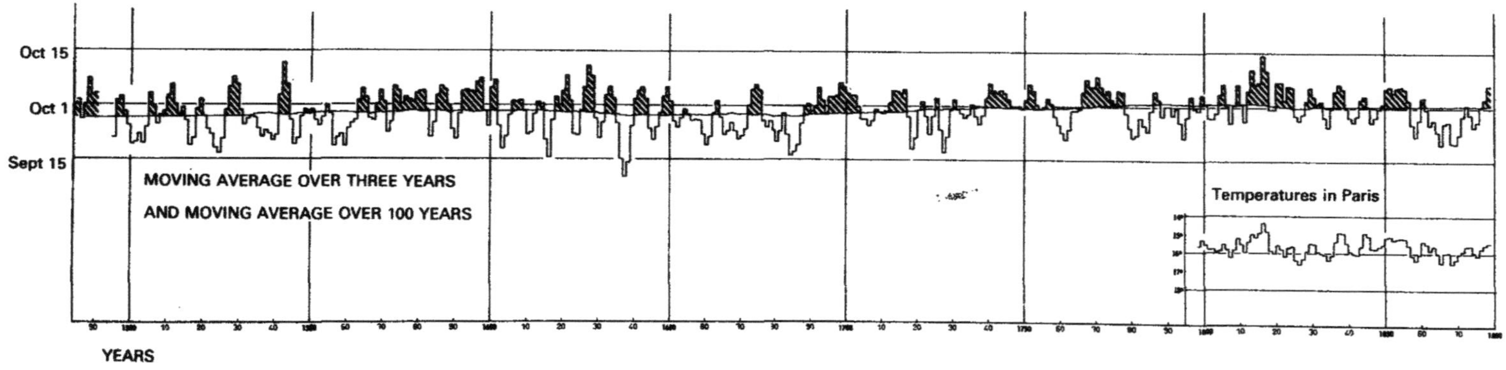

conditions.[5] The English or Dutch temperature series, despite their remarkable similarity to our grape harvest date curve, are too geographically distant from the northern French vineyards to be useful as valid bases for comparison.

In order to make the curve more readable, we have "tempered" it slightly by using a three-term moving average calculated according to the formula:

$$MA = \frac{a + 2b + c}{4}$$

where MA is the moving average, b the base, a the preceding year (preceding b), and c the following year. The final result appears in Figure 1.

Thanks to the use of a large number of base series for the nineteenth century, the most solid period in our construction, we have distinguished between the mainly early grape harvest periods or series of the years and the mainly late ones. The comparison with the thermometric curve indicates that the early phases correspond to the series of warmer spring-summers, and those which are late correspond to the series of cooler spring-summers:

— early series (hot): 1857–1875;
— late series (cool): 1813–1817; 1850–1855.

For the eighteenth century, there are likewise a certain number of grouped years:

— late series: 1713–1716; 1739–1757; 1766–1777
— early series: 1758–1762; 1778–1784.

For the seventeenth century:

— late series: 1640–1644; 1648-1651; 1672–1675; 1688–1698;
— early series: 1636–1638; 1676–1687.

The peaks of lateness should also be noted for 1600, 1601, 1621, and 1627.

For the sixteenth century, we have pointed out in our earlier article the contrast between the slightly early period on the average of 1500-1559 and the late and probably cooler period of

5 Marcel Garnier, "Contribution de la phénologie à l'étude des variations climatiques," *La Météorologie* (Oct.-Dec., 1955), 291-300.

1560-1609; this second part of the sixteenth century coincides towards the end with a general forward advance of the Alpine glaciers during the years 1590-1602. One may surmise that the glaciers had been "fattened up" as the result of the reduction of summer ice ablation due to the slightly accentuated coolness on the average of the summers between 1560 and 1600.[6]

Appendix I

Côte-d'Or: Dijon, Volnay, Beaune
Jura: Salins
Duché de Bade: Kürnbach
Würtemberg: Stuttgart
Cher: Bourges, Lons-le-Saunier
Doubs: Besançon, Vuillafans
Seine-et-Oise: Argenteuil, Montmorency, Clamart, Carrières-sur-Seine, Cormeilles-en-Parisis, Herblay, Houilles, Brétigny-sur-Orge, Leudeville, Marolles, Sucy-en-Brie, Villiers-le-Bel, Juziers, Marcoussis, Grigny, Juvisy, Longjumeau, Morangis, Noisy-sur-Oise, Sartrouville, Cormeilles-en-Vexin, Frémécourt, Marines, Bougival, La Celle-Saint-Cloud, Marly-le-Roy, Port-Marly, Aubergenville, Les Mureaux, Villebon, Conflans-Sainte-Honorine, Poissy, Verneuil-Vernouillet, Boissy-l'Aillerie, Cergy, Neuville-sur-Oise, Jouy-le-Moutiers, Noisy-le-Grand, Chatou, Fourqueux, Le Pecq, Chaville, Garches, Meudon, Fleury, Saint-Cloud, Ville-d'Avray, Viroflay
Seine: Romainville
Seine-et-Marne: Varreddes, Villenoy, Germigny-l'Epine, Trilport
Loiret: Denainvilliers, Boësses, Ladon
Yonne: Auxerre, Avallon
Vosges: Avrainville, Marainville, Villers, Châtillon-sur-Saône
Meurthe-et-Moselle: Sommervillers, Foug
Meuse: Jony-sous-les-Côtes, Verdun
Marne: Cramant, Cauroy-lez-Hermonville
Nièvre: Livry
Loir-et-Cher: Vendôme
Indre: Vatan
Sarthe: Château-du-Loir
Haute-Saône: Gy, Bucey-les-Gy, Morey, Vesoul
Aube: Arsonval, Couvignon, Essoyes, Loches, Les Riceys
Allier: Bresnay, Contigny
Switzerland: Aubonne, Lausanne, Lavaux, Morges, Vevey, Rivaz, Villeneuve, Genève, Le Crest, Rolle et Veytaux.

6 Baulant and Le Roy Ladurie, "Les dates de vendanges au XVIe siècle."

Appendix II- Correlations

For the nineteenth century (1790-1879) we obtained a certain number of excellent correlations between our series.

Correlation coefficients between series:
Côte-d'Or/Franche-Comté: 0.82
Côte-d'Or/Seine, Seine-et-Oise: 0.85
Côte-d'Or/Marne, Seine-et-Marne: 0.83
Côte-d'Or/Meuse, Meurthe-et-Moselle, Vosges: 0.87
Côte-d'Or/Aube, Yonne: 0.92
Côte-d'Or/Sarthe, Loir-et-Cher, Loiret: 0.85
Côte-d'Or/Germany: 0.89
Franche-Comté/Cher, Indre: 0.86
Franche-Comté/Seine, Seine-et-Oise: 0.83
Franche-Comté/Aube, Yonne: 0.86
Franche-Comté/Germany: 0.81
Seine, Seine-et-Oise/Cher, Indre: 0.80
Seine, Seine-et-Oise/Marne, Seine-et-Marne: 0.80
Seine, Seine-et-Oise/Meuse, Meurthe-et-Moselle, Vosges: 0.83
Seine, Seine-et-Oise/Aube, Yonne: 0.90
Seine, Seine-et-Oise/Sarthe, Loir-et-Cher: 0.87
Seine, Seine-et-Oise/Germany: 0.83
Cher, Indre/Aube, Yonne: 0.83
Marne, Seine-et-Marne/Meuse, Meurthe-et-Moselle, Vosges: 0.84
Marne, Seine-et-Marne/Aube, Yonne: 0.84
Marne, Seine-et-Marne/Germany: 0.85
Meuse, Meurthe-et-Moselle, Vosges/Aube, Yonne: 0.90
Meuse, Meurthe-et-Moselle, Vosges/Germany: 0.86
Meuse, Meurthe-et-Moselle, Vosges/Switzerland: 0.80
Aube, Yonne/Sarthe, Loir-et-Cher: 0.88
Aube, Yonne/Germany: 0.88
Sarthe, Loir-et-Cher/Germany: 0.85
Switzerland/Germany: 0.82

For the eighteenth (1690-1810) and seventeenth centuries (1590-1710), the results are clearly of poorer quality. However, there are some good correlations.

Eighteenth Century:
Côte-d'Or/Aube, Yonne: 0.82
Franche-Comté/Marne, Seine-et-Marne: 0.88
Seine, Seine-et-Oise/Marne, Seine-et-Marne: 0.95
Seine, Seine-et-Oise/Aube, Yonne: 0.87
Marne, Seine-et-Marne/Aube, Yonne: 0.82

Seventeenth Century:
Côte-d'Or/Franche-Comté: 0.87
Côte-d'Or/Seine, Seine-et-Oise: 0.80

The existence of relatively low correlation coefficients between certain series should not be concealed in the eighteenth century and especially not in the seventeenth century. For example:

Seventeenth Century:
Côte-d'Or/Switzerland: 0.58

Eighteenth Century:
Seine, Seine-et-Oise/Switzerland: 0.49

On the whole all our correlations between regional series are largely positive.

Barbara Bell

Analysis of Viticultural Data by Cumulative Deviations

Deviations Climatologists wishing to bring out the more lasting and significant shifts in the mean temperature or rainfall from the random annual fluctuations customarily use a running mean of the individual data points. Yet Kraus introduced a method, previously used mainly by hydrologists, to the study of rainfall records; this method consists of computing and plotting the cumulative deviations from a mean, in his case the mean annual rainfall for the years 1881-1940. Although this method, which can be used for any time-series of data, has been adopted to date by very few climatologists, I consider that it has several significant advantages: it accomplishes the purpose of running means by clearly separating brief random fluctuations from more lasting and significant shifts; but it does not obscure the abruptness or gradualness of any significant change; nor does it smooth away extreme events, the exact year of any abrupt change, or the existence and precise duration of clusters of extreme years. All of these last items of information are obscured or lost in the usual method of smoothing the data by the use of running means.[1]

I have computed cumulative deviations from the mean date of grape harvest compiled from historical records by Le Roy Ladurie, and have analyzed the revised series of dates presented in his research note which appears in this issue. The analysis of this most recent data series from Le Roy Ladurie and Baulant is presented below. In addition, the method has been applied to the data on vine yields in the vicinity of Zurich, which underlies Pfister's Figure 3, above. Finally, I have analyzed Le Roy Ladurie's index of wine quality. Although other factors also influence the growth of the vine, Pfister has shown that these three viticultural parameters cast light primarily on the temperature variations of

Barbara Bell is Astronomer at the Harvard-Smithsonian Center for Astrophysics.

1 Eric B. Kraus, "Secular Changes in the Rainfall Regime of SE Australia," *Quarterly Journal of the Royal Meteorological Society*, LXXX (1954), 591-601; *idem*, "Graphs of Cumulative Residuals," *ibid.*, LXXXII (1956), 96-98.

the early, mid-, and late summer, respectively, over their years of record.[2]

The cumulative deviations are obtained by calculating the cumulative summation, year by year, of the individual deviations from a mean, which may be taken over any selected interval within the data set. The results can be expressed either directly in the units of measurement, in units of the standard deviation, or in percentages. To obtain an impression of the statistical significance of any trend in a cumulative-deviation curve, it is useful to express the deviations and the cumulative deviations in units of the standard deviation (S.D.) from the mean, rather than simply in the units of measurement.

DATE OF GRAPE HARVEST Figures 1a and 2a present cumulative deviation graphs for Le Roy Ladurie and Baulant's dates of the grape harvest from Western Europe (the environs of Paris to the Rhineland and Switzerland) from 1484 to 1879, divided into two overlapping data sets. At first sight it might appear more appropriate to present a single curve based on a mean over the entire period of record. However, the 100-year running mean in Figure 2 of Le Roy Ladurie and Baulant shows a gradual retardation in the harvest date, measured in days after September 1. This retardation, occurring at diverse times in different regions during the reign of Louis XIV, is discussed more fully by Le Roy Ladurie in his book. There he interprets it as "a sign of a viticultural and not of a climatic revolution . . . [because] in order for the delay to take on meteorological significance and indicate a long-term climatic movement, it has to occur in all vineyards at once." But it does not do that even within small geographical areas. Since I wish to illustrate climatic, not viticultural changes, I therefore divided the data into two temporal regimes. Trial calculations, based on various intervals for the mean, suggest that in the data

2 Emmanuel Le Roy Ladurie (trans. Barbara Bray), *Times of Feast, Times of Famine: A History of Climate Since the Year 1000* (Garden City, 1971); Le Roy Ladurie and Micheline Baulant, "Grape Harvests from the Fifteenth through the Nineteenth Centuries," in this issue; Christian Pfister, "The Little Ice Age: Thermal and Wetness Indices for Central Europe," in this issue. I wish to thank Pfister for making his table of vine yield data available to me in advance of publication.

Fig. 1 Cumulative Deviations, in Units of Standard Deviation (S.D.).

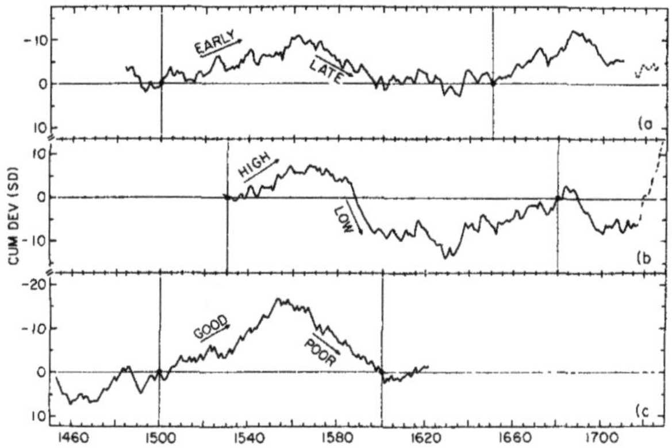

Cumulative Deviations:
a —of dates of grape harvest (Le Roy Ladurie and Baulant, "Grape Harvests," this issue) for 1484–1728 data, based on mean of 150 years, 1501–1650 = 27.43 ± 9.33 days after Sept. 1;
b —of vine yield (Pfister, personal communication) for 1529–1728 data, based on mean of 150 years, 1531–1680 = 18.2 ± 9.5 hectoliters/hectare;
c —of the index of wine quality (after Le Roy Ladurie, *Times of Feast*, 371–375): for 1453–1622 data, based on century mean, 1501–1600 = 0 ± 5.

Fig. 2 Cumulative Deviations, in Units of Standard Deviation (S.D.).

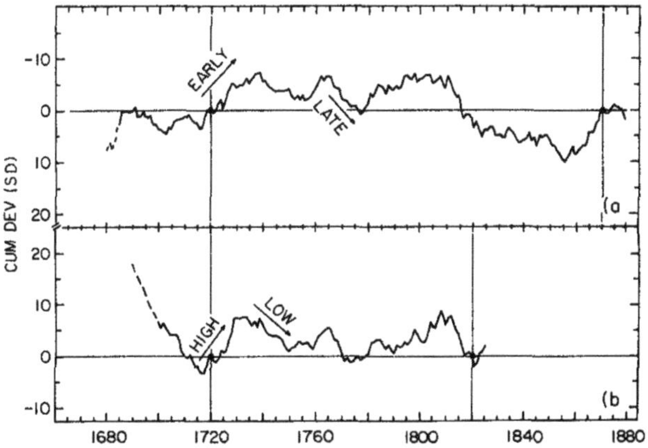

Cumulative Deviations:
a —of dates of wine harvest for 1670–1879 data, based on mean of 150 years, 1721–1870 = 29.74 ± 7.85 days;
b —of vine yield for 1690–1825 data, based on mean of 100 years, 1721–1820 = 26.4 ± 12.8 hl/ha.

normalized to Dijon, the viticultural transition occurred around 1700, a point we shall return to below.[3]

Figure 1a shows the cumulative deviation curve for the dates of wine harvest from 1484 to 1728, based on a mean = 27.43 ± 9.33 days (after September 1) for the 150 years 1501-1650. Very similar appearing curves were obtained using means for 1501-1600 (27.45 ± 9.25), 1501-1700 (27.17 ± 9.06), and 1601-1700 (26.89 ± 8.86). Figure 2a shows the cumulative deviation curve for the dates of wine harvest from 1670 to 1879, based on a mean of 29.74 ± 7.85 days, for the 150 years 1721-1870. Similar appearing curves were obtained using means for 1701-1825 (29.75 ± 7.80) and 1771-1870 (29.93 ± 8.19).

In a plot of cumulative deviations, *the important feature is the slope of the curve* during any time interval. A succession of large positive (negative) deviations from the mean will produce a graph-segment with a steeply rising (declining) slope. I have inverted the curves of Figures 1a and 2a so that a rising curve indicates a period of early harvests (negative deviations), and hence years of relatively warm spring-summer weather; and a declining curve indicates a period of late harvests, and hence years of relatively cool and wet spring-summer weather. Note particularly that the highest points on the graphs do not mark the year(s) of earliest wine harvest; they mark the year(s) of transition from a sequence of early harvests to late harvests. Thus we have a peak at 1561-1563, marking the end of some seven decades of early harvest (with relatively large dispersion) and the onset of some forty years of relatively later harvests and cooler summers during which the Alpine glaciers were presumably growing in preparation for their damaging advances recorded during the 1590s. Other conspicuous transitions occur at 1686, 1740, and around 1765.

Conversely, the lowest points on the graph mark years of reversal from a sequence of late to early harvests. The most notable of these minimum reversals occurred in 1856-1857, a manifestation in the dates of grape harvest of the end of the little ice age and the start of the modern recession of Alpine glaciers. The reversals singled out for mention are abrupt, essentially V-

3 Le Roy Ladurie and Baulant, "Grape Harvests," Fig. 2; Le Roy Ladurie, *Times of Feast*, 242.

shaped, and suggestive of sudden transitions between quasi-stable climate regimes. They reveal important information that would be lost with the use of running means. Centered on 1800 we have an example of a gradual change, terminated by a major abrupt change in 1812-1813. A flat portion of the graph, whether above, below, or along the zero line, represents a period of wine harvest dates close to the mean.

Two features may be noted in support of the division of the data set at 1700. First, it is widely agreed that the 1690s were cool and wet in Europe, so that we should expect a downtrending graph-segment, which appears more strongly in Figure 1a than in 2a. Second, Pfister notes that "suddenly, from 1719 to 1729 the vine became exuberant . . . casks overflowed . . ." from which we should expect warm summers and early harvests and an uptrending graph-segment, which appears strongly in Figure 2a and scarcely at all in 1a. The harvest dates of the 1720s are close to the mean of 1501–1700, and earlier than the 1721–1870 mean. If we infer that the summers of the 1720s were warm and favorable to the vine, then the harvest dates should be earlier than their norm. Only if placed in the second data set, after the viticultural retardation of harvest dates, will they precede their (later) norm, as expected.[4]

VINE YIELD Figures 1b and 2b show cumulative deviation curves for Pfister's vine-yield data, pre- and post-1700 respectively, plotted to facilitate comparison with the curves based on the dates of harvest. Figure 1b shows the cumulative deviation curve for vine yield from 1529 to 1728, based on the mean of 18.2 ± 9.5 hectoliters/hectare for the 150 years 1531-1680. Figure 2b shows the cumulative deviation graph for vine yield from 1690 to 1825, based on the mean of 26.4 ± 12.8 hl/ha for the hundred years 1721-1820. It is noteworthy that after 1700 the mean yield increases substantially, and would indeed dominate the shape of the curve if I had plotted a single graph for the entire period of Pfister's data. In the absence of firm information, I hypothesized a change in vine productivity that was brought about by human cultivators—by improved techniques of cultivation and/or new varieties of grape vines—around 1700 associated with the change

4 Pfister, "Thermal and Wetness Indices," 687.

in harvest date discussed above. Hence I divided Pfister's data set also into two independent segments, Figure 1 pre-1700 and Figure 2 post-1700. Although differing in year-to-year detail, the cumulative deviation curves for vine yield and harvest date vary in parallel in their more important fluctuations. The principal difference between parts (a) and (b) appears in the years of transition between segments, between Figure 1 and 2, especially over the years 1700-1718 when the mid-summers were persistently cool but the late-spring/early-summer fluctuated between warm and cool years. It may be, however, that this lapse from parallelism results more from geographical than from seasonal factors, and/or that the viticultural changes leading to increased yields came into effect after those changes favoring later harvest dates.[5]

WINE QUALITY Figure 1c shows a cumulative deviation plot of the index of wine quality published by Le Roy Ladurie, based on the mean for the years 1501–1600.[6] This curve again differs only in detail from those for harvest dates and vine yield over the same time period—the most notable detail being that the reversal from preponderantly "good" to "poor" wine quality precedes the others by about a decade. It would appear that the major shift to cooler summer temperatures, associated with the onset of the little ice age, occurred first in the late summer, between 1554 and 1559, and second in the early summer in 1563. The shift in mid-

5 Present data do not permit me to distinguish the relative importance of these factors. Pfister (*ibid.* Fig. 3) introduces a linear trend in the mean to adjust for this increase in yield. I chose to use a discontinuity instead, primarily for consistency with the other viticultural parameters analyzed. A graph of cumulative deviations from a mean-with-linear-trend would differ from Fig. 1b primarily in showing a higher relative yield for 1529-1568, and a lower yield for 1640-1684. The underlying assumptions about the evolution of viticulture are quite different. The first posits a uniformly continuing increase in vine productivity underlying the temperature-imposed variations; the second posits a more or less abrupt increase over some years around 1700.

6 If one's primary aim is to compare data sets by the method of cumulative deviations from a mean, one should use a mean over the same interval of years for each data set. This was not my initial aim and, furthermore, proved impractical here because the other graphs were completed before I received Pfister's data. His data began with 1529.

A cumulative-deviation curve always passes through zero at both ends of the mean interval. Since Fig. 1a passes through zero at 1600, as well as at 1500 and 1650, the means of Figs. 1a and 1c are compatible, and I had no hesitation in using the longer period of the mean, 1501-1650, for Fig. 1a. A trial calculation for harvest dates based on a 1531-1680 mean established that the different means used for Figs. 1a and 1b introduce no significant distortions in the pattern of fluctuations, as one could also infer from the small slope of a line connecting the points at 1530 and 1680 on Fig. 1a.

summer temperatures appears to have occurred in two stages, a mild shift around 1560, and a stronger shift in 1585 (making up for its delay with a more stable and pronounced shift) as reflected in the yield of the vine.

The three curves—for harvest date, vine yield, and wine quality—differ in detail from year to year. But overall they show a strong parallelism, rising and falling together with the main shifts in the prevailing temperature regime throughout the summer, from late spring through to early autumn—a parallelism the more remarkable because the data of Le Roy Ladurie and Baulant, and Pfister come from different areas of Western Europe. The basic parallelism of the three curves, however, cannot be regarded as proof that the temperature regimes of early, mid-, and late summer varied to a like degree throughout Western Europe. The month attributed to each parameter is that found by Pfister to have given the highest correlation coefficient between the viticultural parameter and the temperature at Basel. The best correlations were only .46 (April–June, date of harvest) and .54 (July, yield). Thus there is an undoubted, if as yet unquantified, integrative effect on all viticultural parameters of the temperature and its variations throughout the growing season, as well as an influence exerted by rainfall, hail storms, insects, and other phenomena.[7]

With this caution, I present in Table 1 a summary of the principal intervals, or clusters of years, of above- and below-average spring-summer temperatures, derived from Figures 1 and 2. The table identifies precisely the years of change, to facilitate comparison with other data on the climate between 1454 and 1879 and comparison of details among the three types of viticultural data.

The method of cumulative deviations shows its advantage over the method of running means because it enables us to identify precisely such clusters of anomalous years—a point of no little importance since the existence of clusters of anomalous years suggests that the climate commonly changes by means of abrupt shifts from one to another among several quasi-stable climate regimes. For this reason the method of analysis by cumulative deviations merits wider use to supplement if not to replace graphs of running means in future analyses of climatological records.

7 Pfister, "Thermal and Wetness Indices."

Table 1 Clusters of Years Tending to Above-Average and Below-Average Spring-Summer Temperatures.[a]

WARMER INTERVALS			COOLER INTERVALS		
APR.-JUNE	JUNE-AUG.	AUG.-SEPT.	APR.-JUNE	JUNE-AUG.	AUG.-SEPT.
EARLY HARVEST DATES	HIGH VINE YIELDS	GOOD WINE QUALITY	LATE HARVEST DATES	LOW VINE YIELD	POOR WINE QUALITY
					1454-60
		1473-84			
			1488-93	—	1488-92
1500-04		1493-1510			
1515-24		1517-23			
			1527-29	—	1526-29
1530-40	—	1531-41			
			1541-44	1542-44	
1545-62	1545-59	1543-53			
			1563-70		1559-66
			1573-82	1576-79	1568-74
			1585-87	1585-1602[b]	1577-1602[b]
			1591-97[b]		
			1600-01		
		1610-20			
			1616-35	1618-29	—
1636-39	1634-38	—			
			1640-43		
			1648-50	1647-52	
1651-71	1653-72				
			1672-75	1673-75	
1676-86	1677-84				
			1687-1703	1685-1701	
			1713-16	1701-17	
1718-37	1718-30				
			1740-45	1740-50	
1757-62	1760-65				
			1767-77	1766-73	
1778-84	1774-84				
				1786-90	
	1791-98				
				1799-1801	
	1803-08				
			1812-24	1813-21	
	—		1835-38	—	
			1850-56		
1857-70[c]					

a Derived from Figures 1 and 2. The more prolonged anomalies of one sign, with minor interruptions of the opposite sign, are marked by vertical lines.

b Period of first documented widespread damage from growing Alpine glaciers of little ice age Le Roy Ladurie, *Times of Feast*, 140.

c Start of major Alpine glacial recession, end of little ice age.

The Contributors

Reader on Climate and History

REID A. BRYSON is Director of the Institute for Environmental Studies at the University of Wisconsin, Madison. He is the author with Thomas Murray of *Climates of Hunger* (Madison, 1977).

CHRISTINE PADOCH is Assistant Professor at the Institute for Environmental Studies at the University of Wisconsin, Madison.

JAN DE VRIES is Professor of History at the University of California at Berkeley. He is author of *The Dutch Rural Economy in the Golden Age* (New Haven, 1974).

HELMUT E. LANDSBERG is Professor Emeritus in the Institute for Physical Science and Technology at the University of Maryland. He is the author of *Weather and Health* (New York, 1969).

ANDREW B. APPLEBY is Associate Professor of History at San Diego State University and is the author of *Famine in Tudor and Stuart England* (Stanford, 1978).

CHRISTIAN PFISTER is Research Fellow of the Swiss National Science Foundation.

JEROME NAMIAS is Research Meteorologist and Head of the Climate Research Group at Scripps Institution of Oceanography.

DAVID HERLIHY is the Henry Charles Lea Professor of History at Harvard University. He is the author, with Christiane Klapisch, of *Les Toscans et leurs familles* (Paris, 1978).

JOHN D. POST is Associate Professor of History at Northeastern University. He is author of *The Last Great Subsistence Crisis in the Western World* (Baltimore, 1977).

JOHN A. EDDY is Senior Scientist at the High Altitude Observatory, Boulder, Colorado.

THOMPSON WEBB III is Associate Professor of Geological Sciences at Brown University.

HAROLD C. FRITTS is Professor of Dendochronology at the University of Arizona.

G. ROBERT LOFGREN is Research Specialist in the Laboratory of Tree-Ring Research at the University of Arizona.

GEOFFREY A. GORDON is Research Associate in the Laboratory of Tree-Ring Research at the University of Arizona.

ALEXANDER T. WILSON is Professor of Chemistry at the University of Waikato, New Zealand. He is currently Director of Research for the Duval Corporation, Tucson.

DONALD G. BAKER is Professor in the Department of Soil Science at the University of Minnesota.

DAVID HACKETT FISCHER is Warren Professor of History at Brandeis University.

THEODORE K. RABB, co-editor of the *Journal of Interdisciplinary History*, is Professor of History at Princeton University.

EMMANUEL LE ROY LADURIE is Professor of History at the Collège de France.

MICHELINE BAULANT is a member of the Centre National de la Recherche scientifique.

BARBARA BELL is Astronomer at the Harvard-Smithsonian Center for Astrophysics.

GPSR Authorized Representative: Easy Access System Europe - Mustamäe tee 50, 10621 Tallinn, Estonia, gpsr.requests@easproject.com

www.ingramcontent.com/pod-product-compliance
Ingram Content Group UK Ltd.
Pitfield, Milton Keynes, MK11 3LW, UK
UKHW021457290926
475585UK00012B/17

* 9 7 8 0 6 9 1 6 4 2 4 1 3 *